A Student's Dictionary of Psychology

A Student's DICTIONARY OF PSYCHOLOGY

Peter Stratton
Senior Lecturer in Psychology, University of Leeds
and
Nicky Hayes
Hon. Research Fellow, University of Huddersfield

Edward Arnold
A member of the Hodder Headline Group
LONDON NEW YORK MELBOURNE AUCKLAND

Edward Arnold is a division of Hodder Headline PLC
338 Euston Road, London NW1 3BH

First published in the United Kingdom 1988
Second edition 1993

Distributed in the USA by Routledge, Chapman and Hall, Inc.
29 West 35th Street, New York, NY 10001

British Library Cataloguing in Publication Data
Stratton, Peter
Student's Dictionary of Psychology.-
2Rev.ed
I. Title   II. Hayes, Nicky
150.3

ISBN 0 340 56926 3

Library of Congress Cataloging-in-Publication Data
Available on request

10 9 8 7 6 5
99 98 97 96 95 94

Typeset in 11/12 Times Roman by Saxon Printing Ltd, Derby
Printed and bound in the United Kingdom by
J W Arrowsmith Ltd, Bristol

Preface

In this second edition of 'A Student's Dictionary of Psychology' we have kept to our original aims but have tried to improve the text in various ways. Firstly, we have extended it both by adding a substantial number of new entries and cross-references, and by expanding some of the old definitions. Secondly, the rate of progress in some areas of psychology has been so rapid that certain entries have needed updating after only five years, and we have changed some entries to take account of this. And thirdly, we have tried to ensure that the dictionary will be useful for almost anybody studying psychology at an introductory level, whatever form that study takes. While we have continually borne the requirements of the A level psychology student in mind, an ever-increasing range of professional and academic courses include a component of psychology and we hope that this dictionary will be helpful for students who need a sound understanding of a part of psychology without necessarily tackling the whole subject.

In writing this dictionary, our intention is to provide a guide for students, particularly those who are encountering psychology for the first time, and who are likely to be bemused or confused by psychological jargon. Many words are used in psychology which are also used in everyday speech; and yet their psychological meaning is different, either because they are used with a far more precise and/or restricted meaning, or because they have a meaning which is completely different from everyday usage. The term 'schizophrenia' might be a common example of the latter.

When dealing with the major terms we have where possible said something useful about the phenomenon as well as indicating how the term is used in psychology. In some cases, it may seem as though we have produced definitions for words which were perfectly clear already. Terms like 'play' have an adequate meaning in everyday language, but the point of defining them in the dictionary (or anywhere else, like an essay) is that the process of defining forces you to make assumptions explicit and gives an opportunity both to reach a conclusion and to know what the conclusion entails when answering questions like 'do only mammals play?' or 'is there such a thing as personality?'

We have occasionally referred to famous psychologists, or important psychological publications in the text. The full references for these are given in the bibliography at the end. On those occasions where we have been referring to an idea which has been defined somewhere else in the text, the reference has been emphasized, even if it appears in a slightly different form as an entry. Consequently, if you are following up such words and you don't immediately see the word that you're looking for, always look at the adjacent items with similar spelling.

Essay-writing and revision

One of the first tasks in both writing essays and revising for exams involves making sure that you have a clear grasp of the meaning of the terms involved. We see this as an explaining dictionary, which means that our first objective is to provide useful and easily-understood definitions which will explain how these terms are used in the discipline of psychology.

Another of the tasks involved in essay-writing and revision is making sure that you comprehend the full range of issues covered by the topic. We have tried to make sure that the cross-references provided in this dictionary will guide you through the range of connections. For the most part, we have not cross-referenced every possible linking term. Instead, we have indicated in the body of the entry where you should look for useful further information, by putting that word in italics. Essential cross-references which are not mentioned elsewhere in the definition are given at the end. For significant issues, these will guide you to a 'home reference', which will explain the core topic and give you cross references to all the connecting entries.

Finally, both essays and exam answers should tell an interesting and convincing story. The psychology of the human being is a never-ending detective story, and many of our entries are essays in miniature. We hope that you will find them interesting, and that you will want to follow-up the clues that we have provided.

Peter Stratton and Nicky Hayes
May 1992

Acknowledgements

The existence of this dictionary owes much to the support and encouragement which Helga Hanks has so generously supplied throughout its preparation.

We are grateful to the many students and colleagues who commented on draft entries while this dictionary was in preparation, and during the lifetime of the first edition. We would also like to thank John Hyatt and Sally Bradley for their imaginative illustrations (the remaining illustrations were drawn by Nicky Hayes). Finally, we are grateful to Lesley Riddle, of Edward Arnold, for her tireless encouragement and enthusiasm.

A

ABBA design An example of *counterbalancing* of experimental conditions. The first condition (A) is followed by two trials of the second condition (B) then by one of the first. The effect is to average out *order effects* if they are constant. But if all of the *practice effects* would take place during the first trial, it would be safer to run half of the subjects with a BAAB sequence.

An alternative use of the term is to refer to experimental designs in which one group of subjects experiences experimental conditions in the order A then B, while the other group experiences B then A. The two uses can be distinguished by examining the number of trials which the subject is obliged to undergo.

Aberrant Behaviour (in biology, an organism) that deviates from what is *normal*, expected, or desirable.

Ability A capacity or skill.

Ablation The removal or destruction of part or parts of the brain by means of surgical techniques, usually involving the cutting or burning away of the tissue concerned.

Abnormal A term applied to behaviour or people who have been classed as not normal. A potentially controversial label because of problems in defining *normality*. Abnormality can be defined in several different ways: (1) as behaviour which is different from the *norm* (i.e. unusual); (2) as behaviour which does not conform to social demands; (3) as statistically uncommon behaviour, based on the assumptions of the *normal distribution*; (4) as behaviour which is maladaptive or painful for the individual; or (5) as the failure to achieve self-actualization, the humanistic view. These criteria bring their own problems because they lead to the classification of highly regarded individuals like artists and social reformers as 'abnormal'.

Abnormal psychology The psychology of abnormal behaviour. This term has largely been replaced by *clinical psychology* when referring to the professional practice of abnormal psychology.

Abreaction A process used in some forms of *psychotherapy*, especially psychoanalytically oriented ones, which involves the re-living of deep emotional experiences. During abreaction, repressed emotional disturbance is brought to consciousness, allowing a recognition of its existence and an opportunity for the client to develop new coping strategies.

Abscissa The horizontal or x-axis of a graph. See also *ordinate*.

Absolute refractory period The period of a few milliseconds immediately after the firing of a neurone. During the absolute refractory period the neurone will not produce another electrical impulse, no matter how much stimulation it may receive. See also *relative refractory period*.

Absolute threshold The minimum amount of stimulation required for an event to be detected. The absolute threshold of a particular form of *stimulus* is set at the point where 50% of the signals with that physical value are detected.

Abstract thought Thought which uses concepts which do not have an immediate material correspondence, such as justice or freedom. In Piaget's theory of cognitive development, the capacity for abstract thought is only acquired after the age of about 12 years. It is an essential aspect of Piaget's *formal operations* stage.

Abuse (i) Of substances, using them inappropriately in a way that is damaging to the individual, e.g. excessive alcohol consumption; sniffing glue. See *addiction*.

(ii) Inappropriate and harmful treatment of another person. See also *child abuse*.

Accent A distinctive pattern of ways of pronouncing words and phrases which is shared by members of a social or regional group. In some circumstances, accent is taken as an important signifier of social status, and may thus determine the nature of *social interaction* between individuals. This is particularly noticeable in highly stratified societies such as Britain. See also *dialect, psycholinguistics, speech registers.*

Accommodation (i) In biological terms, the process of adjusting shape to fit incoming information; e.g. the process by which the lens of the eye adopts a different shape when the eye is focused on distant objects, than when it is focused on nearby objects.

(ii) In Piagetian theory, the process by which a *schema*, or cognitive structure, becomes adjusted to new information by extending or changing its form, or even by subdividing into a set of schemata with different applications. See also *assimilation, equilibration.*

Account analysis A research method which involves analysing the accounts which people give of their experience. Developed in answer to the need for psychological research techniques which could deal with the subjective realities of human experience as opposed to external behavioural assessments, account analysis takes as its starting point the radical idea that what people say may have meaning. From there it goes on to assert that a systematic approach to collecting people's own versions of an experience or event may be of value to the psychologist seeking to understand human experience. Account analysis can take many forms, but generally involves two stages: (1) a systematic approach to the collection of accounts, generally through *interviews*; and (2) some reflective technique which allows the psychologist to extract ideas, themes or implications from the data, such as *discourse analysis, attributional* analysis, and *thematic qualitative analysis.* Account analysis forms an important part of the *ethogenic* approach to the study of social behaviour propounded by Rom Harré, (e.g. Harré, 1979).

Acetylcholine A *neurotransmitter* which is particularly found at the motor end plate and is therefore involved in muscle action. Some military nerve gases have their effect by the destruction of the enzyme at the motor end plate which breaks down acetylcholine, thus causing it to build up, causing uncontrollable muscle spasm. Other drugs prevent the uptake of acetylcholine at the motor end plate by themselves being picked up at the receptor sites, and thus blocking the uptake of the neurotransmitter: the paralysing poison *curare* operates in this way, and *nicotine* has a partial effect of this kind.

Achievement The successful reaching of a goal. Used particularly to refer to real-life successes and when evaluating a person's life.

Achievement motivation The motivation to accomplish valued goals and to avoid failure. This concept became important as motivation theory became less dominated by physiological drives.

Achievement test A test designed to measure what a person has already achieved; e.g. a statistics exam. See also *aptitude test.*

Achromatic colours A range of hues which is judged to be all of one colour, and which will have wavelengths within a narrow band, although they may vary in intensity and saturation. 'Achromatic' usually means 'all of one colour'.

Acoustic Concerning sound and sound quality.

Acquisition (i) A term used to indicate that a particular skill or ability has been gained by an animal or human being. When applied to language, the term 'acquisition' is used to avoid drawing inferences about whether language has been learned or inherited. Stating that a skill has been acquired implies that the actual process by which the skill was obtained is not the issue being discussed at that particular time.

(ii) The phase during a *conditioning* procedure in which the response is learned or strengthened.

Acronym An abbreviation of a title by giving the initials of each word, particularly common in discussions of psychological tests, e.g. BAS for British Ability Scale. Groups sometimes develop acronyms that outsiders do not understand as a way of excluding non-members and producing a feeling of cohesion.

Acting out To express a wish, need or motivation, particularly when it is unrecognized or unconscious, in overt behaviour. Often the behaviour is aggressive and self-destructive and may be very uncharacteristic for the person, who may have no idea why they behaved in that way.

Action pattern See *fixed action pattern*.

Action potential The electrical impulse produced by a neurone when its stimulation exceeds the threshold level, such that the neurone fires. See also *evoked potential*.

Action research Action research is an approach to psychological enquiry which challenges the idealised view of the psychologist as an 'objective' scientist, standing apart from the subject matter and observing it dispassionately. Instead, action research takes as its starting point the idea that the presence of the researcher will always affect behaviour, and so it is naive to assume that the activities of the researcher will have no effect on the behaviour of the subject. Instead, an action researcher will deliberately act as a change agent within a given situation, and incorporate the effects of these actions as an integral part of the outcome of the research. Initially developed in an organisational context by Lewin (e.g. Lewin, 1947), action research has continued to be popular in organisational psychology. With the increased emphasis on *ecological validity* in psychological research, action research is gradually gaining acceptance in several other areas of psychological investigation. See also *new paradigm research, participant observation*.

Actualizing tendency A term coined by Rogers (1954) to describe the process by which people seek to develop their various potentials and to maximize their personal growth, once their need for positive regard from others has been satisfied. See also '*self-actualization*'.

Acuity The fineness of the discrimination that a sense organ can make. Most commonly used of vision, where visual acuity indicates the smallest objects that can be distinguished.

Adaptation The process of adjusting to an environment in such a way that maximal benefit may be obtained from it, or at least in such a way that life may be continued in a reasonably productive manner. The term has highly specific meanings in: (1) physiology: the adjustment of bodily organs to particular environmental demands, e.g. the adaptation of the heart to living at a high altitude; (2) evolutionary biology: how a species is matched to the environments in which it has developed; and (3) psychology: the process by which an individual achieves the best balance feasible

3

between conflicting demands. Piaget uses the term more specifically for the processes by which *cognitive* structures are made to correspond to reality. See *accommodation, assimilation.*

Addiction A state of physiological or psychological dependence on some substance, usually a drug, resulting in tolerance of the drug such that progressively larger doses are required to obtain the same effect. Addictions are most clearly identified by a failure to function adequately when the substance is withdrawn (see *withdrawal symptoms*). The commonest addictions are for socially accepted drugs such as nicotine and alcohol, though illegal drugs (e.g. heroin) and those initially taken as medical treatment (e.g. *tranquillizers*) often cause more public concern. Treatments have covered the full range of psychological and psychiatric techniques, but behavioural and group methods are most widely used. Colloquially, the term has been stretched to cover needs which have become exaggerated to a degree that is damaging to the individual, e.g. 'addiction' to television, violent exercise, or food. See also *dependency.*

Adipose Fatty, or pertaining to fat. Adipose tissue in the body is tissue which stores fat; adipocytes are cells specifically adapted for that purpose.

Adjustment Originally, adjustment was regarded as little more than the avoidance of *maladjustment*, but then became a goal for therapy with the emergence of the *humanistic* approaches to psychotherapy. Modern therapists accept that many forms of adjustment are possible, thus avoiding value-judgements about life-style. Broadly speaking, adjustment refers to the individual's achieving a harmonious balance with the demands of both environment and cognitions. The development of behavioural technologies to improve individual adjustment raises complex ethical considerations, e.g. whether conditioning techniques to solve problems of sexual adjustment can be adopted without consideration of values and morals.

Adolescence The developmental period between childhood and adulthood. In some cultures the transition is very brief and achieved through some form of *rite of passage* but in Western culture it extends from the onset of *puberty* around 12 years of age, to about 17 or 18. Research on adolescence has tended to emphasize the four developmental areas of competence, *individuation, identity* and *self-esteem.*

Adrenaline A hormone and neurotransmitter, produced by the adrenal glands, which is particularly associated with emotional states. Adrenaline is involved in states of arousal, initiated by the action of the sympathetic division of the *autonomic nervous system.* It is released as hormone by the adrenal gland, and serves to maintain an activated state of the body, such that a higher level of energy is produced by the autonomic functions. It also acts within the brain as a neurotransmitter, and again is involved in emotional states. Adrenergic pathways or fibres are those in which stimulation results in the production of adrenaline. Sometimes it is called epinephrine, especially in America.

Aesthetics The study of the nature of beauty, or of pleasing perceptual experiences.

Aetiology The study of causation. This term is particularly used to refer to the causes of illnesses and mental disorders.

Affect (i) A term used to mean emotion, but covering a very much wider band of feeling than the normal emotions. Affect includes pleasurable sensation, friendliness and warmth, pensiveness, and mild dislike etc, as well as the extreme emotions such as joy, exhilaration, fear and hatred. Broadly speaking, affect refers to any category of feeling, as distinct from cognition or behaviour.

(ii) As a verb: to influence; to have an *effect*. Note that the verb 'to *effect*' means 'to cause'. 'An *effect*' is a result.

Affect display A set of physical changes which indicates an emotional state. Examples are the *pilomotor response* in cats, indicating fear, and the greeting smile in humans, indicating friendliness.

Affectionless psychopathy A term used by J. Bowlby to describe a syndrome in which an individual does not demonstrate any emotion, whether positive or negative, towards any other human being. Affectionless *psychopaths* were characterized by a lack of social conscience and a high level of delinquency.

Affective disorder A psychiatric term used to refer to syndromes in which the patient appears to be producing inappropriate emotional responses. Alternatively, a prolonged disturbance of mood or emotion, as in mania and *depression*.

Affective domain (i) A traditional approach to understanding human personality, originating with the ancient Greeks, involving seeing the psyche as comprising three domains: the *cognitive* domain, housing the intellect and reason, the *conative* domain, concerning the will and intentionality and the affective domain, housing the emotions and feelings. The traditional model is of a charioteer driving two horses: the charioteer represents the cognitive domain, steering the chariot and planning the route, while the motive power behind human action is provided by the *conative* and affective domains.

(ii) This distinction is maintained in attitude theory, in which an attitude is considered to consist of three major components: a cognitive component, consisting of justifications for the attitude, an affective component involving the emotional aspects of the attitude and a behavioural component, concerning the tendency to act in accordance with the attitude (which corresponds to the conative domain).

Afferent neurone A nerve cell (neurone) which carries information in the form of electrical impulses from the sense organs to the central nervous system. See also *sensory neurone, efferent neurone*.

Affiliation The process of joining or the sense of belonging to a *group*. Nearly everybody feels a desire to belong, so affiliation has been treated as a need or motive.

Afterimage An image which remains in the visual field after the original stimulation has ceased. Afterimages usually occur after particularly intense or prolonged stimulation of the retina. See also *negative aftereffects*.

Age regression See *regression*.

Agentic state See *autonomous state*.

Aggression A term used in several ways, commonly to describe a deliberate attempt to harm another being. There is no agreed definition, partly because the term is sometimes applied to behaviour (hitting), sometimes to an emotional state (feeling aggressive), and sometimes to an intention (wanting to harm). There are several classifications of different kinds of aggression, the most useful distinction being between *instrumental aggression*: an aggressive act performed in order to achieve some other objective, and *hostile aggression*: motivated by antagonistic feelings and emotions.

Agoraphobia The commonest form of *phobia*. Literally a fear of open spaces, it is usually associated with a fear of interacting with other people. Agoraphobia results in a severe restriction of the sufferer's life, as they cannot enter any crowded area and may become unable to leave the house. Often is it possible to recognize some way in which this makes it unnecessary for them to have to tackle some source of anxiety. Psychological treatments may attempt either to reduce the symptoms of the phobia by such techniques as *systematic desensitization*, or to resolve the underlying anxiety.

Aha! experience A sudden experience of enlightenment, in which the solution to a problem is perceived very rapidly, with little prior feeling that progress is being made towards the solution. An aspect of insight learning, utilized by *Gestalt* theorists such as Kohler to argue against the reductionist approach to human learning put forward by the *behaviourist* school. See also *creativity*.

Alarm reaction A term used to describe the series of physiological responses brought about by the activation of the sympathetic division of the *autonomic nervous system*. Investigated systematically by W. Cannon, the alarm reaction involves, among other changes, increased heart rate and blood pressure producing an increased supply of oxygen to the muscles; changes to the digestive system producing rapid digestion of sugars for increased energy, and alterations to the composition of the blood such that clotting occurs more quickly. The effective result of these changes is that the body is prepared for extended and demanding effort.

Alcoholic A person who has become dependent on the drug alcohol. Many problems dealt with by clinical psychologists are caused or aggravated by alcohol, e.g. some 30% of cases of *child physical abuse*. Alcoholism is treated in a number of different ways by different practitioners, including clinical psychologists, with varying degrees of success. There is controversy over the question of whether total abstinence is essential for anyone who has been an alcoholic. Alcoholism is probably the most widespread and damaging *addiction*. See also *antabuse*.

Alienation A state of feeling separated from (i) oneself and one's own feelings; or (ii) other people and society.

Alienist An early title for psychiatrists, who treated 'aliens' (lunatics).

Algorithm A technique used for solving a problem which will eventually lead to the correct solution if undertaken systematically and consistently. Algorithms generally work by progressively reducing error until the goal is reached. Many problems cannot be solved by this method, and some can be solved more quickly by more *insightful* techniques. See also *heuristic*.

Allele One form of a paired *gene*. Most organisms have pairs of *chromosomes*, with matching genes situated on each chromosome. If the two alleles are different in form, one may be *dominant* over the other (e.g. in eye-colour, brown is dominant over blue), or both may contribute to the eventual phenotype (e.g. in skin colour, a combination of the two characteristics results). Partial dominance is also possible.

All-or-none principle The principle that a neurone either fires or it does not, with no variation in the strength of the electrical impulse. It was originally thought that all nerve cells operated according to the all-or-none principle: implying a necessity for digital processing models of brain functioning, and fostering some *computer simulation* approaches to understanding cognition. However, more recent evidence has shown that all-or-none firing is uncommon within the brain itself, and the cortical neurones may use variable coding.

Alpha male A term used in *ethology* to describe a top-ranking or dominant male in a social group. See *dominance hierarchy*.

Alpha waves Distinctive patterns of brain activity shown on EEG readings, having a wave pattern of between 8 and 12Hz, characteristic of a state of wakeful relaxation.

Altered states of awareness Also known as altered states of consciousness, this is the idea that there are qualitatively different mental states which will result in various psychological processes such as *attention* and *motivation* functioning differently. Sleep is an obvious example, but more subtle changes in the waking state have also been studied.

Altruism Acting for the benefit of other people without regard to personal cost or benefit. There is dispute about whether truly altruistic behaviour ever occurs. See also *reciprocal altruism.*

Alzheimer's syndrome A condition which resembles *senile dementia* but which occurs much earlier in life, with some sufferers even being as young as 40 years old.

Ambiguous Having more than one possible meaning. An ambiguous stimulus is one which can be interpreted in more than one way. [f.]

Ambivalence The simultaneous existence of two opposed emotions, motivations or attitudes, e.g. love–hate; approach–avoidance. Each feeling has its own separate

An ambiguous figure which may be seen as a letter or a number

origin, so the two cannot be reconciled and the person either alternates between the two attitudes or *represses* one of them.

Ambivert A person who has achieved a balance between extreme *introversion* and extreme *extraversion*, as described by Eysenck.

Ameslan A standardized sign language used by deaf-and/or-dumb people in America. Several primate studies have involved the teaching of Ameslan to gorillas or chimpanzees, with a degree of success.

Ames room A well-known *visual illusion* in which a room is constructed which, when viewed from a particular viewing point, appears to be normal, but which in reality has one corner very much farther away from the viewer than the other. The appearance of equal distance is achieved by carefully balancing the perspectives of the room and the levels of the floor and ceiling. The effect is that people or objects of the same size appear to be of different sizes. [f.]

Amnesia Loss of memory, normally from physical causes. Retrograde amnesia refers to loss of memory for events prior to the damaging event or disease: loss of memory

Amphetamine

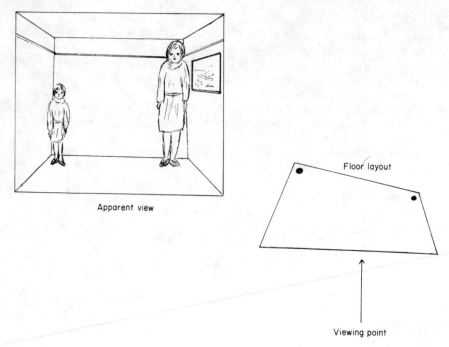

Apparent view

Floor layout

Viewing point

The Ames room illusion

of the few minutes leading up to severe concussion is the most common example. Anterograde amnesia refers to the loss of subsequent memory: for instance, impairment of the ability to code new memories after brain surgery, or, as found in *Korsakoff's syndrome,* through long-term alcoholism.

Amphetamine A drug which stimulates the central nervous system. It is usually prescribed in order to raise energy levels or to prevent sleep; and abused (as 'speed' etc.) for the same purposes. Amphetamine is also used as an appetite suppressant to help dieters and to control *hyperactive* children.

Anaclitic depression A depression caused in infants between 6 and 18 months by prolonged separation from their mothers. The term was first used by Rene Spitz, and was an important concept in early studies of *maternal deprivation.*

Anagram A puzzle or problem which consists of words with their constituent letters disarranged, such that all the necessary letters are present but in the wrong order. The letters may be randomly listed (GAANMRA) or rearranged to resemble other words (A GRANMA). Anagrams are often used in laboratory problem solving tasks.

Anal stage The second of Freud's psychosexual stages, in which *libido* focuses on the anus. See also *oral stage, phallic stage.*

Analogue An object or phenomenon which corresponds to another in at least some respects. The term is used in: (1) theories of memory referring to information stored in the brain from which a representation or image of an object can be generated; (2) in biology for characteristics of different species which have the same functions; and (3) in electronics for information stored through a continuously variable quantity. Compare analogue (circular) clock faces as opposed to *digital* watches.

Analysand That which is being, or has been, analysed. Used particularly to refer to student analysts undergoing *psychoanalysis* as part of their training.

Analysis (i) Identifying the constituent parts or links of a whole so that it can be better understood and interpreted, e.g. in statistical analysis.

(ii) A shorthand term for *psychoanalysis*.

Analysis-by-synthesis A term given to a cognitive model in which the brain is seen as combining separate pieces of information about an event in order to make the best judgement about the nature of the event.

Analysis of variance (ANOVA) A statistical procedure to test whether groups of scores differ from each other. The principle is that if the scores are not being influenced in different ways, the variation (*variance*) of scores within each group will allow us to predict how much variation there will be between the means of the groups. If it turns out that the group means vary more than expected, we conclude that the groups differ (and have therefore been influenced in different ways). Several different sources of influence can be tested within a single ANOVA design, and the complex relationships or *interactions* between them can be analysed.

Analytic psychology The system of psychopathology and treatment devised by Carl Jung after his split from the Freudian school. It introduces concepts such as the *archetype* and the collective unconscious.

Androgens Hormones produced mainly by the *testes*. They are responsible for the physical developments in the foetus which give rise to male characteristics including the external genitalia. Later in life they influence sexual activity, and the expression of genetically controlled characteristics such as the growth of a beard. See *testosterone*.

Androgyny The presence in one person (either male or female) of both male and female characteristics. In humans, there are no *sex differences* which are present in one gender and not the other: it is more a matter of the prevalence and strength of each tendency. Therefore, everybody mixes male and female characteristics to some extent, and the term androgyny is reserved for people who show both male and female characteristics to a significant degree. Research indicates that individuals who are psychologically androgynous tend to be mentally healthier than those who represent orthodox gender stereotypes.

Anecdotal evidence Information quoted in support of an idea or theory which has been obtained purely from everyday experience or accounts, rather than from some form of systematic or controlled study.

Anencephalic Without a *cerebrum*. Anencephalic infants usually survive only for a few days after birth, although some have been kept alive for up to six months. Anencephalic infants are of interest to students of *neonate* functioning, as observable difference between them and normal infants only seem to emerge after the first few weeks, implying that *cortical* activity may not play an important part in early infant behaviour.

Angst A mental disquiet or anguish considered by *existentialists* to be the inevitable outcome of a full appreciation of the implications of personal responsibility and personal choice.

Angular gyrus That part of the *cerebral cortex* which is involved in the decoding of visual symbols. The angular gyrus receives input from the *visual cortex*, and appears to process that information into a form equivalent to information which has been processed by the *auditory cortex*. The angular gyrus then passes messages on to the area known as *Wernicke's area*, where it is processed for comprehension.

Accordingly, the angular gyrus plays an important role in the process of reading, and it is thought that damage to this area is the root cause of certain *dyslexias*.

Animism The attribution of living qualities to inanimate objects or phenomena; and frequently the attribution of conscious awareness. Animism is a powerful trend in human thought processes which has been studied mostly in the thinking of young children. It is commonplace in everyday speech, e.g. referring to the family car as a person, and occurs extensively in the belief systems of non-technological cultures.

Anodyne A pain-relieving treatment or agent.

Anomaly A noticeable deviation from what is expected or predicted.

Anorexia nervosa A disorder in which the person becomes unable to eat and may starve to death. Anorexia is most common among teenage girls, often initiated by excessive dieting. Anorexia has been thought of variously as arising from a distorted body image; a subconscious attempt to return to pre-pubertal physique and, by implication, social role; and as an expression of rebellion against domination by a mother figure. See also *bulimia, eating disorders*.

ANOVA See *analysis of variance*.

ANOVA model of attributions See *covariance*.

Anoxia A reduced supply of oxygen to the brain or other tissues. Particularly likely to happen to a baby around birth, and can result in brain damage.

ANS See *autonomic nervous system*.

Antabuse The commercial name for the drug disulfiram, which produces an extreme reaction when taken in conjunction with alcohol. Usually administered by a skin implant which can last for a month or more, antabuse is used therapeutically in *aversion therapy* for *alcoholics*. The association between extreme nausea and vomiting, and alcohol can sometimes produce a lasting aversion to alcohol, enabling the alcoholic to deal with the problem.

Antagonistic Having an opposite effect, working against or competing with something else. Antagonistic muscles work in opposite ways to one another, e.g. one set of muscles in the iris contracts to dilate the pupil of the eye, while a different set contracts to constrict it.

Antecedent Taking place before the relevant event. An antecedent may be the cause of the event, but it cannot be assumed that it was.

Antecedent variables Factors in an experiment which precede (happen before) some other event. Because of the time relationship, the antecedent variable cannot have been caused by the subsequent event, and may even have been a cause of it.

Antenatal The period before birth.

Anterograde amnesia Loss of memory for events taking place after the damage producing the amnesia. See also *amnesia, retrograde amnesia*.

Anthropology The study of different human societies, involving a particular emphasis on social structures and belief systems. An anthropologist is one who undertakes such a study, often using non-participant observational techniques.

Anthropomorphism The attribution of human qualities such as personality, emotions and motives to animals.

Anticipation A mental state of readiness for a specific event. See also *set*.

Anticipatory schema A concept put forward by Neisser, whereby an anticipatory schema forms an essential *cognitive* component in the cyclic process of *perception*. An anticipatory schema consists of a set of cognitions derived from the individual's beliefs and experiences, based on observations of the situation, and concerning the most probable outcomes of action. This schema will be utilized in the selection of appropriate behaviours. The actions thus directed will modify the situation, producing a new sample of observations. These in turn will serve to modify the anticipatory schema. Neisser considered this continuous cyclic process to be the key to an understanding of human cognition. [f.]

Anxiety A stressful state resulting from the anticipation of danger. Anxiety has a physiological component (the *alarm reaction* or fight or flight reaction), a cognitive

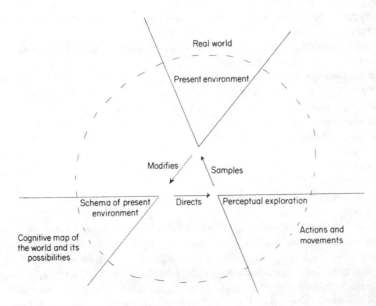

Neisser's model of the perceptual cycle

aspect, particularly in narrowing attention, and a subjective experience of discomfort. Each of these components may help the person deal effectively with clearly recognized, real and immediate dangers, but can be damaging both psychologically and physically when the anxiety persists, as in occupational *stress* or unresolved *unconscious* conflicts.

Anxiety disorder General term for psychological disorders in which chronic anxiety, with debilitating consequences, is a prominent feature. Includes *panic attacks* and *phobias*.

Apathy A mental state characterized by lack of interest in everyday or momentous events, and involving a disinclination to exertion or effort in order to effect or achieve results of any kind. The state is characteristic of *depression*.

Aphagia A lack of eating, which can be experimentally induced by lesions in the lateral *hypothalamus*. Animals with aphagia show no interest in solid food, to the point of starvation. Aphagia is believed by some physiological psychologists (though not all) to be a mechanism in *anorexia nervosa*.

Aphasia A disorder of speaking, sometimes brought about by *lesions* in *Broca's area* – the area of the cortex involved in speech production. The individual has serious problems in articulating words, though no difficulties in understanding language.

Apnoea (apnea) Temporary stopping of breathing, common among premature babies. Present in some adults during sleep and believed to result in the destruction of brain cells through *anoxia* in extreme cases. Often associated with snoring.

Apparent motion A term used to describe visual illusions which provide an appearance of movement even when no such movement is actually occurring. Examples of this are found in the *phi phenomenon*, the *waterfall effect*, and *stroboscopic* stimuli.

Apperception The conscious awareness of an act of perception, with focused concentration on its full meaning.

Appetitive behaviour Behaviour which is directed towards the satisfaction of some kind of desire, want, or need.

Applied psychology A general term used to classify areas of psychology in which theories are put to use in dealing with practical, non-laboratory situations. Applied psychology traditionally includes *clinical psychology, educational psychology, industrial* and *occupational psychology*; but also includes other fields where psychological theories may be put to use, such as *environmental psychology* or *study skills*.

Apprehension (i) In colloquial terms, a feeling of unease or dread concerning some future event.

(ii) In cognitive terms, the mental grasping or full comprehension of a concept or idea.

Approach–avoidance conflict A pattern of behaviour often seen when an organism is inclined or required to approach something which has simultaneously attractive and aversive qualities – e.g. a parachute jump. The individual tends to oscillate between approach behaviour and avoidance behaviour, with approach behaviour typically dominant when the event or stimulus is more distant in time or space, and avoidance becoming more characteristic when the event or stimulus is closer.

Apraxia Disorders of movement or of the control of movement which have been caused by damage to higher brain centres rather than by failures of sensory feedback or muscle control.

Aptitude The ease with which a person will acquire a new set of skills or abilities. An individual is said to have an aptitude for a particular skill if she learns that skill more rapidly and with more ease than other individuals with the same prior knowledge of it.

Aptitude test A test to assess the ease with which a person will acquire specified skills, i.e. a measure of aptitude for some kind of competence. See also *attainment test*.

Archetypes Classic, powerful images which, according to Carl Jung, are held in the *collective unconscious* and recur frequently in folk art and mythology. Examples of Jungian archetypes are: the earth mother, the sea as a symbol of rebirth, the omnipotent father, the inaccessible virgin, the knave, etc.

Arousal A state in which the *sympathetic division* of the *autonomic nervous system* is activated, producing an *alarm reaction*, or a longer term response to *stress*. Arousal is characterized by very high levels of adrenaline in the bloodstream, and results in a general state of readiness to react in the organism. Depending on cognitive and environmental factors, this may result in anger, anxiety, exhilaration, excitement, or, if the arousal is frequent and prolonged and the energy is not dissipated by regular demanding exercise, in long-term stress disorders.

Arrhythmic Irregular, lacking in rhythm.

Articulation (i) Clear verbal expression.

(ii) Free movement through the action of a joint, sometimes extended to mean the assembly of joints and levers that make such movement possible, e.g. in *robotics*.

Artificial insemination (AI) The introduction of sperm into the vagina or uterus of a female by technical means rather than by sexual union. As the donor of the sperm may be unknown to mother and child the technique has implications for family relationships and the possible selection of genetic characteristics.

Artificial intelligence (AI) An area of research which aims to develop computer systems which will allow the computer to develop novel solutions to problems, or to produce other forms of 'intelligent' behaviour such as gathering relevant information to aid expert decisions. Computer systems which can 'reason', it is hoped by those involved, will eventually be able to produce the same kinds of outcomes as are produced by human cognitive processes. Work on artificial intelligence has tended to concentrate on: (1) knowledge-based systems, known as 'expert systems', which are capable of limited decision-making on the basis of prior input from a number of human experts; (2) man–machine interface research, such as the development of *voice recognition systems*; and (3) *robotics*, the development of sensing and manipulation devices such as might be suitable for manufacturing processes. See also *computer simulation, parallel distributed processing*.

Asch effect A term used to describe conformity arising through awareness that, if the individual stated their own judgement, they would be responding differently from the rest of the group, and that others would be aware of that dissent. Asch's studies of *conformity* involved a subject placed in a situation in which the other group members had been primed to give obviously wrong answers to a relatively simple problem and the real subject had to answer openly, after the majority had answered. [f.]

Assertion training A series of *therapeutic* techniques designed to enable the individual to take an active or dominant role in social interaction.

Assimilation One of two processes by which a *schema* in Piagetian theory is considered to develop. New information is said to have been *assimilated* when it is fitted into an existing schema and so can be understood in relation to earlier learning. Assimilation and *accommodation* are considered to be continuous cognitive processes, contributing to the generalized process of cognitive *adaptation*. See also *equilibration*.

Association The linking of one thing with another in sequence. Associative learning is learning which has been acquired as a result of the connection of a *stimulus* with a *response*. During the period when psychology was attempting to account for *all* behaviours as stimulus–response connections, association was seen as the central psychological process.

Association cortex A term given to those parts of the cerebral cortex which do not seem to have a specific, localized, function. They have been thought of as the areas in which basic perceptual information is associated with more general knowledge. See also *equipotentiality*.

Assortative mating The tendency for organisms (including humans) to select as sexual partners those with characteristics similar to their own.

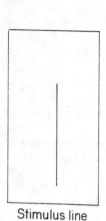
Stimulus line

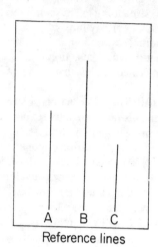
A   B   C
Reference lines

The test material used in Asch's study of conformity

Assumption An idea or set of ideas which is taken for granted in the formulation of an argument or theory. If made explicit, it is sometimes called an axiom.

Astigmatism A disorder of vision in which lines at certain angles are not perceived with the same *acuity* as lines of different orientations. Recent research suggests that astigmatism may arise from the irregular development of *simple* or *complex cells* in the *lateral geniculate nuclei* of the *thalamus* – involved in the primary decoding of visual information.

Asymmetrical Lacking in balance or evenness.

Attachment A close, emotionally meaningful relationship between two people in which each seeks closeness with the other and feels more secure in their presence. The attachment between mother and infant has been extensively studied, and some writers apply the term only to the relationship of the infant to the mother. Attachment has been the subject of much theorizing by John Bowlby. There is now much evidence that the quality of attachments in infancy affects *exploration* and *play* in the short term, and a wide range of cognitive and social functions throughout childhood. However, it is no longer believed that the infant always forms a major attachment exclusively to the mother. See also imprinting, monotropy, bonding, contingency, strange situation technique.

Attainment test A test designed to assess the knowledge and skills which an individual has obtained, either through experience or through following a prescribed course of training. See also *aptitude test*.

Attention A directed focus of awareness, such that the individual is maximally prepared to respond to a specific kind of signal or sensory input. See *selective attention*.

Attenuation (i) The shortening or limiting of an object or event.

(ii) A term used by Triesman (1964) to refer to the weakening of a signal being processed, as an essential part of a model of selective attention. See *filter models*.

Attitude A 'mental set' held by an individual which affects the ways that that person responds to events and organizes their cognitions. Attitudes are commonly held to have three essential components or dimensions: a *cognitive* dimension, involving the beliefs and rationalizations which 'explain' the holding of the attitude; an *affective* dimension, involving the emotional aspects of the attitude, such as likes, dislikes, feelings of distaste, or affection; and a conative, or behavioural, dimension which involves the extent to which the individual is prepared to act on the attitude that they hold. See also *prejudice, stereotype.*

Attribution The process by which an event or the behaviour of a person is understood in terms of suggested motives or influences. The judgement that a person or an event has a particular characteristic or quality.

Attribution theory An extensive and growing area of social psychology dealing with the ways that people attempt to account for their own and other people's behaviour. It is most concerned with the kinds of causes by which people come to account for their experiences – attributions about negative *life events* are considered to be particularly important. Attribution theory has been used to improve and extend *helplessness theory* and is extensively used in *cognitive therapy.* Strictly, attribution theory deals with how people come to have their beliefs about the causes of events and behaviour; while attributional theory deals with the different forms (or *attributional styles*) that such beliefs may take. See also *covariance, social attribution.*

Attributional error The universal tendency to see one's own behaviour (particularly when it has undesirable consequences) as a rational response to the situation, and other people's behaviour as originating in their characters. So, when I crash the car, it is because of poor visibility and an icy road; but I attribute my friend's crash to the fact that he is careless and impatient. This is also known as the fundamental attributional error.

Attributional style The theory that individuals tend to believe in particular kinds of causes for a wide range of effects. Styles may vary in the extent to which they incline towards *stable* causes (ones which are unlikely to change in the future), *global* causes (affecting lots of things), and *internal–external* causes (such as character or situation). So, of two people who have failed an exam, one may attribute the cause to the room being noisy (unstable, specific and external), while the other may believe it is due to their being stupid (stable, global and internal). Martin Seligman and others have produced evidence that individuals who incline towards using a stable, global and internal pattern of attributions may be particularly vulnerable to *depression.* See also *personal, controllable.*

Audience effects The effects produced by the presence of other people, on the individual's behaviour. See also *social facilitation.*

Audition The process of hearing. Auditory signals are processed by means of a complex auditory system: the outer ear collects the signals and focuses them inward; the middle ear amplifies the signals; the inner ear transduces the signals into electrical impulses; the auditory nerve transmits the signals to the brain via a cross-over junction with the auditory nerve from the other ear. Some primary decoding of the signals occurs in a region of the *thalamus* and they are eventually interpreted in the *auditory cortex* of the *cerebrum.*

Auditory cortex That part of the *cerebral cortex* involved in the interpretation of sensory messages received by hearing. The auditory cortex is located on the *temporal lobe* of the *cerebrum,* immediately below the *lateral fissure.*

Authoritarian personality A specific, rigid pattern of personality characterized by punitive approaches to social sanctions and high levels of prejudice towards out-group members. Adorno showed that the cognitive styles of highly prejudiced right-wing conservatives had two distinctive traits: (1) rigidity – maintaining a belief system even in the face of direct evidence showing that it is untrue or inefficient; and (2) intolerance of ambiguity – a tendency to take sides quickly and to be unable to cope with equivocal positions. Adorno concluded that this was due to defence mechanisms: highly prejudiced individuals had to protect themselves against ambiguities which might challenge their ideas. Also, they had often been brought up by cold and highly authoritarian parents, producing a reaction formation; the child would displace its aggression towards authority figures onto minority groups in society.

Adorno developed the F-scale (F for fascism), which measured authoritarianism through nine sub-traits. These were: (1) conventionalism; (2) authoritarian submissiveness; (3) authoritarian aggression (hostility towards those who challenge authority); (4) anti-intraception (a tough-minded punitive approach); (5) superstition and stereotype (a belief that events are externally controlled rather than controllable by the individual); (6) power and 'toughness' (a tendency to behave in a dominating manner); (7) destructiveness and cynicism; (8) projectivity (a tendency to project unconscious impulses onto others); and (9) sex (an exaggerated concern with sexual misbehaviour).

Authoritative A term used by Baumrind to describe a style of parenting or *child rearing* in which children are encouraged to participate in decision-making and to express their opinions, but the parent nonetheless has the final authority. This was in contrast with an *authoritarian* approach, in which the child is not encouraged to express an opinion; or a *laissez-faire* approach in which the parent has little involvement in the process of decision-making.

Authority figure A person who represents power or established dominance in some way.

Autism (i) Thought and fantasy determined entirely by the person's needs and wishes and not constrained by reality in any way. Daydreams are autistic, but the term is usually reserved for the more extreme and permanent removal from reality of *schizophrenic* thought.

(ii) A serious disorder appearing towards the end of infancy, in which the child withdraws from all social contact, which seems to be aversive and distressing. Activity is directed towards inanimate objects and may give evidence of quite high intelligence, but speech is usually minimal. Although it is often called infantile autism, or childhood autism, the condition can persist throughout the person's life. There is little agreement about cause, although a majority of those who work in the area probably believe in an organic predisposition, and even less agreement about treatment.

Autochthonous A term used to describe a state arising primarily from events within the individual – such as thirst, or hunger.

Autogenic Originating from the self; self-initiated, e.g. autogenic training in which the individual is trained to have internal control of their own relaxation.

Autohypnosis *Hypnosis* which has been self-induced. Many forms of hypnotherapy concentrate on the development of the individual's own skills in autohypnosis, so that they can develop strategies for *coping* with stressful events.

Autokinetic effect A visual illusion involving the apparent motion of a stationary dot of light when it is perceived in a totally dark environment. The light appears to move in rapid jerks.

Automatic routines Actions or sequences of actions which have become so habitual that we no longer need to pay attention to them. Much complex skill learning consists of developing automatic routines (such as changing gear while driving). One of the most powerful demonstrations of automatic routines in cognition is the *Stroop effect*, which shows how the automatic subroutine of reading conflicts with the visual identification of colours.

Automatic writing Writing that is performed without conscious awareness by the writer. It is usually elicited under *hypnosis*, but it can be produced by sitting undisturbed for a long period and writing continuously with no attempt to control what is produced. After several hours the product may give an uncensored glimpse into the unconscious or it may not.

Autonomic nervous system (ANS) A network of unmyelinated nerve fibres running from the brainstem and spinal cord to the viscera, which can activate the body rapidly, preparing it for action. The ANS has two main parts, the sympathetic and the parasympathetic divisions. Activation of the *sympathetic division* results in the body being rapidly prepared for action, producing the *alarm reaction*. It is strongly involved in active emotions, such as excitement, fear, or anger. Activation of the *parasympathetic* division involves more quiescent functioning like digestion, tissue growth and repair, the storage of blood sugars and the building up of bodily reserves. The parasympathetic division is thought to be involved in the passive emotions, such as depression, contentment, or sadness. There seem to be individual differences in the balance between sympathetic and parasympathetic arousal, and in the overall lability of the autonomic nervous system. [f.]

Autonomous morality The third of Kohlberg's three stages of *moral development*, in which the individual is considered to have reached a point where she arrives at moral judgements and decisions on the basis of her own reasoning, rather than simply by accepting the ideas laid down by society. In the first level of this stage, the individual accepts social rules and moral codes because she considers them to have been democratically established for the common good; in the second level a more individual judgement is achieved, and the person may eventually come to reject some commonly accepted social values which she feels to be unjust or immoral.

Autonomous state A reference to the theory of obedience outlined by Milgram (1973) in which he proposed that human beings, as social animals, have two alternative and mutually exclusive 'states' of social being. One of these is known as the autonomous state, in which people act as autonomous individuals, and in which personal conscience and morality therefore inform and direct their choices of action. The other, the agentic state, occurs when people act simply as an agent for others higher up in the social order. In the agentic state, individual conscience and morality become suppressed, as the individual is no longer acting autonomously. This, according to Milgram, is the mechanism which permits people to engage in behaviour when obeying others which they would find abhorrent when acting in their own private, personal capacity.

Autonomy A state of independence and self-determination in the individual, considered to be the ultimate goal of *humanistic* and *existentialist* therapies.

Average (i) A colloquial term used to mean 'usual' or 'commonplace'.

(ii) An everyday term used to describe a statistical measure of *central tendency*, such as the arithmetic mean.

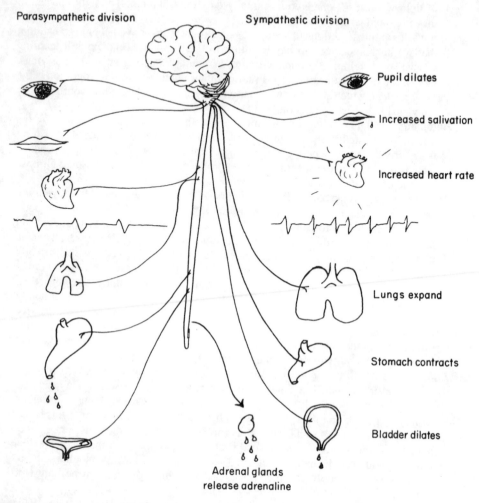

Parasympathetic division   Sympathetic division

Pupil dilates

Increased salivation

Increased heart rate

Lungs expand

Stomach contracts

Bladder dilates

Adrenal glands
release adrenaline

Some connections of the autonomic nervous system

Aversion therapy A technique of *behaviour therapy* which involves utilizing the framework of *classical conditioning* to develop alternative behaviour patterns in the individual. It is done by the association of unpleasant stimuli or consequences with the maladaptive behaviour, such that the behaviour comes to be avoided. See also *antabuse*.

Aversive Leading to avoidance behaviour. A stimulus or event which is unpleasant (such as an electrical shock) would be described as aversive.

Avoidance learning The training of behaviour through a process of *negative reinforcement*, such that an *aversive stimulus* fails to take place if the behaviour is demonstrated. Avoidance learning is extremely resistant to extinction.

Awareness A subjective state of being alert or conscious: cognisant of information received from the immediate environment. See also *altered states of awareness*.

Axon The elongated 'stem' of a neurone, by means of which the electrical impulse is passed from one region of the nervous system to another. The axon is that part of the neurone found after the cell body, taking the direction of travel of the impulse. In *afferent* or *sensory neurones*, the elongated part of the neurone found before the cell body is known as the *dendron*.

B

Babbling Vocalizations produced by infants, which include the full range of human phonemes. In 'Verbal Behaviour' Skinner argued that *language acquisition* occurred as a result of behaviour shaping, with infant babbling as the *operants, conditioned* through the *law of effect*.

Babinski reflex A *reflex* of newborn babies in which scratching the sole of the foot produces extension of the toes. Absence of the reflex indicates a damaged motor system in a baby. Conversely, organic damage is shown if the reflex is present in an adult.

Baby talk The style of speech adopted by adults when talking to a baby, also called 'motherese'.

Backward conditioning A variant of *classical conditioning* in which the unconditioned stimulus (UCS) precedes the conditioned stimulus (CS). There is not yet agreement over whether backward conditioning is possible. If it can occur, it is certainly difficult to achieve. See also *trace conditioning, simultaneous conditioning, delayed conditioning*.

Balance theory A theory put forward by Heider, suggesting that we need to maintain a state of cognitive equilibrium between the different attitudes that we hold, and that our social cognitions would, if necessary, become modified in order to create or perpetuate such a balance. *Cognitive dissonance* is a later variant of the theory.

Balanced design An experimental design in which sources of variation such as practice, fatigue or sex of subjects, are balanced so that they will not be responsible for differences between the groups. See *ABBA*.

Balanced scale A test or questionnaire in which sources of bias in the items are counterbalanced. For example half of the items should be true and half false, so that any tendency to prefer to answer 'yes' does not distort the outcome.

Bandwagon effect The tendency that all people have, to believe a claim or hold an attitude if they believe that most other members of their group have that belief. See also *Asch effect, Barnum effect*.

Barbiturate A widely used drug, particularly given to promote sleep, and to control epilepsy. Barbiturates are highly addictive and commonly abused, producing *amnesiac* disorders in long-term users.

Barnum effect An effect named after the circus entrepreneur T.P. Barnum, whose motto in dealing with the gullible public was 'there's a fool born every minute'. Used to describe the widespread acceptance of certain common beliefs, e.g. astrological predictions which are written in such general terms that they can be readily applied to anyone, but which are read by the credulous as being an exact description of their own individual character or circumstance. In cognitive terms, it refers to the tendency for people to engage in *selective perception*, noticing only what they wish to believe and ignoring that which does not accord with their expectations.

Basal age On tests graded by age, the highest age level up to which all of the items are passed. May be called 'basal mental age' in *intelligence testing*.

Base rate or baseline The level or frequency at which a function is operating before any experimental or therapeutic procedures have been started. Measures taken before an intervention is started may be used as a prediction of what the level of functioning would have been without the intervention. So baseline heart rate may be measured for a few seconds before a stimulus and the recordings used to show whether the heart rate after the stimulus is consistently different; or number of cigarettes smoked per day may be recorded for a month before treatment starts, to see whether there is any change.

Basic needs The most compelling human needs such as food and the avoidance of pain. In Maslow's theory these are at the base of a *hierarchy of needs* and other requirements, even for physical safety, will be ignored until they are satisfied.

Basic trust The development in an infant of total trust that the mother will provide for, protect, and not harm the infant. It is the first of Erikson's eight stages of lifespan development, and is proposed as the most important task that the infant must complete. It is achieved as a result of the security provided by good mothering. Erikson also pointed out that a capacity for mistrust is sometimes useful.

Basilar membrane A membrane running the length of the cochlea, in the inner ear, on which are located hair cells which effect the transduction of auditory vibrations into electrical impulses, for transmission to the *auditory cortex* by the *auditory nerve*.

Battered baby A term coined by C. Henry Kempe in 1962 in a paper which first alerted the medical profession to the widespread existence of infants who had been injured by their parents. See also *child abuse*.

Bayley infant development scales Measures of infant development which assess infants on mental and motor tasks. First developed in the 1920s based on the work of Gesell, but still the most widely used infant assessment. The norms are based on normal infants and rely heavily on the ability of the infant to perform motor tasks, but the scale is now used almost exclusively to assess the general development of children with motor impairments.

Behaviour The movements or actions which a person or animal performs. If something is referred to as 'behavioural', it means that it is only concerned with actual behaviour, and not, for instance, with any cognitive aspects of a performance.

Behaviour disorder A general term used to cover a wide range of psychological disorders in which the behaviour of the person is the major concern. More specifically it applies to conditions such as *psychopathy, addictions,* and *hyperactivity*. One feature of behaviour disorders is that they usually involve symptoms which are likely to bring the sufferer into conflict with society. See *conduct disorder*.

Behaviour genetics The study of the influence of genes on behaviour. Empirical work is concerned particularly with changes of inherited tendencies as a result of selection pressure, and on environmental influences on the expression of these tendencies. Practical considerations result in much of the work being done with organisms with very short breeding cycles, such as fruit flies (drosophila). Applied behaviour genetics has a history of a few thousand years in, for example, horse breeding. See also *eugenics*.

Behaviour modification The therapeutic technique of treating psychological disturbances by dealing solely with the maladaptive behaviour which they produce. The process of behaviour modification operates from the assumptions that disturbed behaviour consists of inappropriate responses to stimuli, arising from maladaptive learning and that new responses may be acquired as a result of new learning. The therapy is based on *conditioning* techniques. Some researchers use the term behaviour modification to refer to those forms of treatment based on *operant conditioning* and *imitative* learning (e.g. *token economy*), and use the term behaviour therapy to refer to techniques based on *classical conditioning* (e.g. *aversion therapy*).

Behaviour shaping The production of novel behaviours through the systematic adjustment of *reinforcement contingencies*. In other words, by rewarding simple behaviours until they are established in the organism's repertoire of actions, and then rewarding only those variants of it which produce behaviour which is even closer to the desired outcome. Once that in turn is established as a frequent behaviour pattern, only behaviour which is even closer to the desired outcome will be rewarded. Behaviour shaping can be used to produce behaviours which are completely unlike anything in the organism's previous repertoire, such as pigeons playing table tennis.

Behaviour therapy The process of treating abnormal behaviour by using conditioning techniques to modify maladaptive symptoms. Behaviour therapy includes techniques such as *aversion therapy, systematic desensitization*, and *implosion therapy*. See also *behaviour modification*.

Behavioural correlates of attention The changes in behaviour or physiological state which people show when they are attending to something, such as turning one's head towards the source of a sound. See also *orienting reflex*.

Behavioural sciences A general term used for those sciences which are concerned with the understanding of behaviour, such as *psychology, ethology*, population genetics, etc.

Behaviourism The school of thought first established by J.B. Watson, in 1913, in which he argued that, to be truly scientific, it was necessary for psychology to concern itself only with that which could be directly observed: the behaviour of organisms. Watson considered that eventually a complete understanding of human behaviour would be developed through the reductionist analysis of psychological phenomena as complex chains of learned stimulus–response connections. The behaviourist approach, developed especially in America and Britain in the first half of the century, proposed that only the study of measurable behaviour was objective and therefore scientific; and that therefore psychologists should study only behaviour and ignore 'mental' processes. Behaviourists also considered that all human behaviour ultimately consisted of links between a stimulus and a response in much the same way as living matter consists of cells. This inherently *reductionist* argument led to much criticism of the approach, which eventually resulted in a considerable decline in popularity. Behaviourist assumptions, however, have left their mark on accepted methodology within psychology, and have formed the background against which *new paradigm research* needs to be seen and evaluated. See also *stimulus–response, reductionism, association*.

Belladonna A drug, atropine, made from the plant belladonna, which dilates the pupils of the eye. Since pupil dilation is a significant non-verbal signal indicating interpersonal attraction, the drug was used as a cosmetic, especially in Italy – hence the name, belladonna: beautiful lady.

Benign Used of conditions which do not pose any significant threat. Opposite is 'malignant'.

Benzodiazepines A group of commonly prescribed minor *tranquillizers* such as Valium. They are a form of barbiturate and produce muscle relaxation, decreased anxiety, and sedation, and are widely used to help people cope with transient difficulties. Benzodiazepines are not *anti-depressants* and there is concern that their ready availability may cause people to put up with an unsatisfactory situation instead of taking positive action to deal with their problems.

Beta rhythm A wave pattern observed in *electroencephalogram* recordings characteristic of an alert, wide-awake individual. See also *alpha waves, delta waves*.

Between-group variance A measure of the variation found among the means of a number of samples. The measure is divided by the within-groups variance to give an F-ratio. These measures are usually computed within an *analysis of variance*. See also *variance*.

Biased sample An error in the way that a particular *sample* has been selected, which results in that sample not being representative of the population as a whole. The classic example is a survey of American electors that was conducted by randomly selecting respondents from the phone book, thereby excluding all of the voters who could not afford phones.

Bilateral transfer The demonstration of a skill learned by one side of the body (e.g. the right hand), by the other half (demonstrated by the left hand). Bilateral transfer can be demonstrated with many motor skills: practice with one side will produce an improvement in performance by the other side.

Bimodal distribution A set of scores which, when plotted as a frequency distribution, shows two separate peaks. Usually this indicates that the scores do not all come from the same population, though it may mean that the source population itself is bimodal. [f.]

Binaural cue An indication of the direction of a source of sound arising from differences in the sounds reaching the two ears. For example, a sound which originates to one side will reach the ear on that side marginally sooner than the other

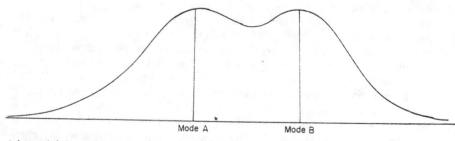

Mode A   Mode B

A bimodal distribution

ear. A cross-over point for the *auditory nerve* means that the difference can be detected, and the direction of sound analysed.

Binocular depth cues Indicators of distance which depend on the use of both eyes. The main binocular depth cues are binocular disparity and the convergence of eye muscles (convergence is greater for objects closer to the eyes). See also *depth perception.*

Binocular disparity The difference in the retinal image received by the two eyes. The disparity between two images is greater for objects close to the eyes and this is used to judge the distance of nearby objects.

Binomial distribution This applies to data which have two mutually exclusive outcomes (heads/tails, adult/child), and it indicates the proportion of each ratio of scores which would be expected for each sample size. So if a coin is repeatedly tossed twenty times, the distribution will indicate how often you would expect to get 20 heads, how often 19 heads and 1 tail, and so on. The actual proportions obtained can be compared with the expected proportions so that it is possible to see whether it is reasonable to suppose that the scores came from the specified population (in this example, whether the coin is unbiased). With large samples the binomial distribution becomes very similar to a *normal distribution.* [f.]

Biofeedback A term used to describe a process by which control of autonomic functioning can be learned if the individual is provided with information about how the body is working, e.g. blood pressure or galvanic skin response readings. Typically, the individual engages in *relaxation* exercises while being provided with such feedback, and it has been demonstrated that effective reduction of blood pressure and heart rate may be achieved in this way. Biofeedback is sometimes cited as an example of the practical application of *operant conditioning*, although this has been disputed on the grounds that the reward – knowledge of results – is a cognitive rather than a behavioural *reinforcement.*

Biogenic A term applied to behaviours or characteristics with a biological origin.

Biogenic amine The group of *amines* which are known to be particularly important in the functioning of the nervous system. Includes *catecholamine* and histamine.

Biological clock The idea that organisms contain a mechanism which maintains a fairly constant rate and which is responsible for controlling biological rhythms such as the sleep/wake cycle. See *biorhythm, circadian rhythm.*

Biological determinism The argument that human nature or human characteristics arise as an inevitable consequence of human biological characteristics. (See also *reductionism*).

Biological therapy The treatment of psychological disturbance or mental illness by physical methods such as drugs, brain surgery, and *electroconvulsive therapy.*

Biopsychology The study of the biological sources of individual functioning. The term usually has a slightly different emphasis to *psychobiology* but there is no universally agreed meaning for either label.

Biorhythm General swings or fluctuations of biological functioning, such as *circadiam rhythms* or the *menstrual cycle.* Longer term biological rhythms, such as annual variations have been demonstrated in many animals, but evidence for their existence in human beings is not yet established. The term has also been adopted by an industry which claims to calculate variations in functions such as creativity from the individual's date of birth. See *Barnum effect.*

Bipolar constructs or concepts The claim that in human thought, concepts are defined in terms of opposite poles. So the concept of honesty entails the opposite

Bipolar depression

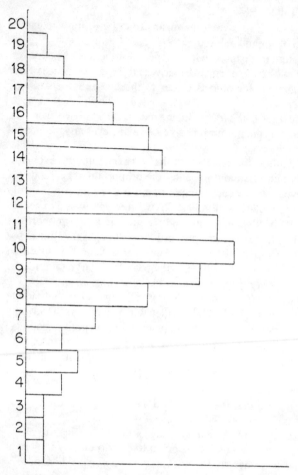

Frequencies obtained by taking successive samples from a binomial distribution.

pole of dishonesty. The most elaborate theory constructed on this basis is George Kelly's *personal construct theory*.

Bipolar depression A disorder of *affect* which involves swings between two extremes of mania and depression. The condition is regarded as having a *biogenic* origin and the swings can be halted by continuous treatment with *lithium*. It is also called manic depression or manic-depressive psychosis.

Birth cry A reflex cry which signals the start of breathing immediately after birth. It is possible for breathing to start without a birth cry.

Birth rate The annual number of live births per 1000 people in the population.

Birth trauma An attempt to explain psychological disturbance as resulting from the trauma of being born. Proposed by Otto Rank in the early days of psychoanalysis but was largely abandoned. Revived more recently in relation to concern about the technological nature of current methods of managing birth.

Bisexual A person whose sexual preference includes people of the same sex as well as those of the other sex.

Bit A term used in *information theory* to define a unit of information. A bit of information is not a vague amount but is precisely defined as the amount required to choose between two equal alternatives: it halves the uncertainty. So if you were searching for a randomly chosen word in this dictionary, one bit would tell you which half it was in, two bits would narrow it down to a quarter, and three bits to an eighth. Twelve bits would identify a specific word out of 4096. The word 'bit' is an abbreviation of 'binary digit'.

Black box A term used to describe an approach to psychological theory in which the internal workings of the organism are regarded as unknowable, as if they take place inside a black box. One is left with the options of either: (1) guessing what is going on in the box by observing the relationships between inputs to the box and its consequent behaviour; or (2) claiming that it is not important to know what goes on in the box, and that only the relationships between input and behaviour should be studied. The second approach was the one chosen by the *behaviourists*.

Blind spot A specific location on the *retina* of the eye where the neural fibres of the ganglion cells in the retina bunch together to form the optic nerve. The blind spot is so named because this part of the retina contains no photosensitive cells, but it is not normally noticed because the brain 'fills in' the area such that it appears to be continuous with the general background.

Blood–brain barrier A characteristic of the blood supply to the brain that prevents many substances from passing from the blood to the brain tissues. It protects the brain from many poisons, but also prevents some potentially useful drugs from being used.

Blood pressure The force with which blood travels through the arteries and veins of the body. High blood pressure is a reliable indicator of long-term *stress*, and a precursor to many stress disorders.

Body image The idea that each individual has of what their body is like. There is evidence of a physiological basis for a body image at birth, but an infant must learn which parts of the universe are not part of its own body. Later the body image extends beyond a representation of the body and comes to reflect an evaluation of bodily characteristics. The 'normal' pattern is to overestimate such characteristics as head size and attractiveness. The body image is an important part of the *self image*.

Body language A general term used to describe those aspects of non-verbal communication (NVC) which involve direct use of the body, such as *gesture, posture* and *proxemics*.

Body-schema The body-schema is the internal representation which the individual has of their own body. According to Piaget, the very first *schema* formed by the infant develops from the first 'me – not me' distinction. For the older person it includes ideas and memories of how the body is, has been, and could be.

Bonding The formation of a strong relationship (*attachment*), usually applied to mothers and their infants, and usually during the period immediately following birth. Some claim that a strong bond may be formed at first contact between mother and baby, a view that has been called the 'Araldite theory' of bonding. Obstetric practices in many Western hospitals have been changed to help foster bonding, but the significance of contact between mother and baby immediately following birth is still a matter of controversy. Some writers reserve the term 'bonding' for the mother's feelings for her baby, and 'attachment' for the infant's relationship to the mother. This usage assumes that there are two, different, one-way processes rather than a *transactional* shared relationship.

Borderline disorder A disorder of personality. The term was originally applied to people judged to be on the borderline between *neurosis* and *psychosis*, particularly those believed to have an underlying psychotic disorder but who were coping reasonably well. It is now used much more broadly for people with instability in their emotions and interpersonal relationships but whose symptoms do not fit any diagnostic system.

Bottom-up processing Perceptual processing which is initiated by the characteristics of the stimulus and leads on to higher forms of cognitive activity, as opposed to top-down processing which begins from the higher levels.

Brain A general term to describe the complex of neural structures developed at the forward end of the spinal cord. In casual usage, however, many psychologists refer to the 'brain' when in fact they mean the cerebrum, or the cerebral cortex (e.g. 'split-brain studies'). Whether the whole brain or simply the cerebrum is meant must be deduced from the context.

Brainstem See *medulla*.

Brainstorming A technique for developing new ideas, commonly used in advertising work and other problem-solving situations. A group undertakes a period of intensive concentration in which any idea at all that comes to mind – regardless of how apparently inappropriate it might be – is noted. There is an agreement not to reject or ridicule any suggestion. At the end of the period of time, all the ideas thus generated are examined for their potential value as a solution to the problem in hand. Some recent research indicates that a group will produce more ideas if the individuals work on their own and then pool their suggestions.

Brainwashing The technique of operating total control over a person's environment, with a consistent application of deprivation, debilitation and dread (the three Ds), so that the victim becomes amenable to adopting a completely new belief system or ideology. The process may depend on some form of *identification*.

Brain waves Overall electrical activity of the brain which can be detected outside the skull by an *electroencephalogram*.

Brightness The intensity of light stimulation or the degree of whiteness in a colour.

Brightness constancy The perceptual adjustment by which we perceive objects seen in widely varying light intensities as being of similar brightness. Brightness *constancy* arises because of the capacity of our perceptual system to work in context, and to deal with relative differences in intensity rather than with absolute ones. So we perceive a piece of paper in the dark coal cellar as brighter than a piece of coal in daylight even though the latter reflects more light.

Broca's area The area of the cerebral cortex at the base of the frontal lobe, usually on the left hemisphere, which is mainly concerned with the production of speech and the formulation of words. Damage to Broca's area can produce *aphasia*.

Bulimia or bulaemia A disorder of eating involving phases in which very large quantities of food may be consumed but which are followed by vomiting, taking laxatives, or intense exercise. The victim therefore gets little nutritional value from the food and may lose weight rapidly. Bulimia is regarded as closely related to *anorexia*.

Bystander apathy A rather moralistic label applied by social psychologists to the phenomenon that onlookers fail to help in emergencies even though they may be upset by what is happening. Concern about bystander apathy was aroused by the case of Kitty Genovese who was stabbed to death in New York in 1964. About 40 people heard her screams for half an hour but none of them even called the police.

Much research has been conducted on the factors that determine whether onlookers will intervene, and the area has come to be called *bystander intervention*.

Bystander intervention The involvement of onlookers in situations where, for example, help is required by another person. The likelihood of bystander intervention has been shown to depend on several factors, such as the onlooker's definition of the situation, the presence of other people who might be expected to provide the help needed, and, to a lesser extent, the characteristics of the victim. The most powerful factor influencing the decision whether to intervene appears to be *diffusion of responsibility*.

C

CA See *chronological age*. Or *child abuse*.

CAL See *computer-assisted learning*.

Cannabis See *marijuana*.

Cannon–Bard theory A theory of emotion put forward in the 1920s, in which it was stated that the psychological experience of emotion, and the physiological reactions produced by the body (see *autonomic nervous system*) were completely independent of one another. Compare *James–Lange theory, interactionism*, and see also *alarm reaction*.

Cardiac muscle Heart muscle. The term 'cardiac' refers to the heart.

Caregiver/caretaker A general term given to refer to the person who looks after a child, thus avoiding the assumptions inherent in the use of terms like 'mother' or 'parent', and allowing for a wider range of possibilities. Despite the apparent opposite, the two terms are used with identical meaning.

Carpentered environment An environment in which there are many straight lines and right angles, e.g. in modern buildings. Carpentered environments are highly characteristic of Western society, and this has been used by some researchers (e.g. Gregory) as a possible explanation for some cultural differences in perception, e.g. that geometric illusions are perceived more or less strongly by people from different cultures. [f.]

Case history A detailed account of the background and previous experience of a single patient or client, which may be important in therapy or in the understanding of a particular psychological phenomenon, such as anterograde amnesia.

Case study A psychological study involving the detailed investigation of just one particular case, or individual. Case studies are extremely important in many areas of psychology, as they allow for an in-depth analysis of unusual circumstances and their outcomes, which in turn may throw light on more usual psychological events (e.g. the outcome of localized brain damage may serve to highlight the functions of a particular area of the brain). They are also used in situations where a detailed account rather than a limited set of standardized measures is required. However the method has its own difficulties, like subjective decisions about which aspects to describe and difficulties of *replication*.

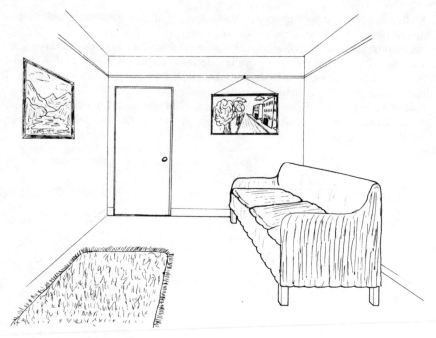

A carpentered environment

Castration threat anxiety A Freudian concept, referring to the anxiety experienced by the young boy during the *Oedipus complex*. As the young boy's sexual interest is directed towards his mother, and his father is perceived as a rival for the mother's love, the child develops a fear that the father (being bigger and more powerful than he) may deal with the competition by castrating him. See also *phallic stage*.

Catastrophe theory A mathematical theory which deals with changes of state which are sudden, substantial, and not easily reversible. Like walking off a cliff. Many psychological phenomena look like this, with examples ranging from spontaneous reversals of perception of a *Necker cube*, through experiences of insight (*aha!*), to the sudden onset of a *phobia*. It is always difficult to record significant psychological phenomena in a form that can be entered into a mathematical equation, and we do not yet know whether catastrophe theory will be useful to psychology.

Catecholamines A group of *biogenic amines* including adrenaline and dopamine, which play a part in neural transmission in the brain. It is suspected that an excess of catecholamines may be involved in *schizophrenia*.

Catharsis The sudden release of tension or anxiety resulting from the process of uncovering repressed trauma or ideas during psychoanalysis. In a wider context, the process of catharsis is seen as the satisfying release of built-up emotional energy, often through *displacement*, e.g. enthusiastic support of team sports.

Cathexis A term used in *psychoanalytic theory* to refer to the investment of intense energy, desire, or meaning in a person, object or event. In many ways cathexis can be thought of as being the opposite of *catharsis*.

Caudal To do with the tail. See *cephalo-caudal*.

Causal attribution A reason given for why an event or characteristic occurred. See *attribution*.

Ceiling effect An effect when a test is too easy so that all of the subjects score near the top (or ceiling) of the scale. The result is that the test is unable to distinguish between individuals who are more, or less, competent. The opposite is known as a 'floor effect'.

Centile The point on a scale such that a given percentage of the relevant population would score at or below that point. So if the 60th centile for height in a given population is 1.75 metres, 60% of the people will be this height or shorter.

Central fissure Also known as the *central sulcus* or the fissure of Rolando, this is a deep groove which runs from the top of the *cerebrum* downwards, separating the frontal and parietal lobes of the cerebrum. The *motor projection area* is located on the frontal edge of the central fissure, and the *somatosensory projection area* is located along the parietal edge.

Central nervous system (CNS) The general name given to the network of nerve fibres and supporting cells which form the brain and the spinal cord. The central nervous system co-ordinates and regulates the major functions of the body, and operates with the other systems of the body, such as the *endocrine system* and the *autonomic nervous system* to maintain integration and effective functioning of bodily and cerebral processes. [f.]

Central sulcus See *central fissure*.

Central tendency See *measures of central tendency*.

Centration A *Piagetian* term which refers to the pre-operational child's tendency to focus on one central characteristic of a problem, to the exclusion of other features. For example judging the volume of a jar of liquid purely by a single dimension such

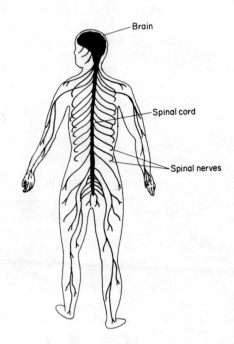

Major pathways of the Central Nervous System

as height, rather than taking into account other dimensions such as width. Centration is considered by Piagetians to be a manifestation of *egocentricity*, which can lead to the inability to *decentre*, and the inability to *conserve* number and volume.

Cephalo-caudal Literally 'from head to tail'. A description applied to the development of motor co-ordination in infants by Gesell, who undertook some of the first systematic observations of infant development, and who proposed that infant development was largely *maturational*, and therefore always consistent in direction. See also *proximo-distal*.

Cerebellum A large, cauliflower-like structure at the back of the brain, which is responsible for the mediation of voluntary movement and balance. The cerebellum is highly convoluted, and has two distinct lobes. It receives sensory input from the *kinaesthetic* nerve fibres and from the organs of balance in the inner ear, and co-ordinates actions into smooth sequences of behaviour.

Cerebral cortex The outer part of the *cerebrum* which has six or seven layers of neurones, and which covers the whole of the surface. The cerebral cortex consists of *grey matter*, and it is in the cortex itself that the information-processing functions of the cerebrum are considered to take place. The remainder of the cerebrum, below the cortex, consists of *white matter*, which is made up of *myelinated* nerve fibres transmitting information from one part of the brain to another. Parts of the cerebral cortex have highly localized functions, such as the *language areas* or the *sensory projection areas*, but large areas of it appear to have a generalized information-processing function, and are referred to by the term *association cortex*. Such areas are considered to conform to the principle of *equipotentiality*, which is to say that

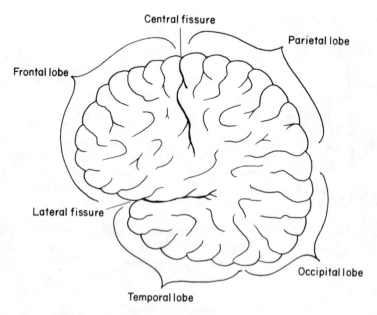

Lobes of the cerebrum (left hemisphere)

they operate en masse, and that overall the amount of cortex involved in functioning is more important than the specific location of that cortex. See also *cerebral hemispheres, cerebrum*. [f.]

Cerebral hemispheres The two halves of the cerebrum, which are joined by a band of nerve fibres known as the *corpus callosum*. In general, the left hemisphere mediates the functioning of the right side of the body, while the right hemisphere is concerned with the left side. Various studies (e.g. Sperry 1967) indicate that the two cerebral hemispheres, while similar in their general structure and in the *projection areas*, are concerned with different aspects of 'higher' mental functioning: the left hemisphere being more concerned with language, logic and mathematical functioning, while the right hemisphere is thought to be more concerned with spatial, artistic and musical abilities. See also *cerebral cortex, cerebrum*.

Cerebrum A structure which in humans forms the largest part of the mass of the brain. The cerebrum is concerned with the processing of information and the co-ordination of voluntary responses, and as such it is also concerned with thinking and other cognitive functions. The cerebrum is divided into two major hemispheres, and each of these has specific areas which deal with localized functions, such as the *sensory projection areas*. Anatomically, each hemisphere of the cerebrum is divided into four lobes: the *frontal lobe*, the *parietal lobe*, the *temporal lobe*, and the *occipital lobe*. See also *cerebral cortex, cerebral hemispheres*.

Chemotherapy The treatment of a disorder or clinical problem by means of drugs. Chemotherapy became a popular method of treatment for psychiatric disturbances during the 1950s, when psychoactive drugs such as *chlorpromazine* (largactil) and diazepam (valium) were introduced. The stronger drugs allowed the treatment of persons with extreme behavioural disorders to proceed without physical restraint, and as such became widely used very quickly. In modern days the spectrum of psychoactive drugs available is extremely wide, including such groups as *antidepressants, anti-anxiety drugs, tranquillizers, amphetamines, barbiturates*, and several more. There is considerable debate as to the ethics and usefulness of many forms of chemotherapy for psychiatric or psychological disorder.

Child abuse The significant failure of a responsible person to care for a child appropriately. Physical injury, sometimes called *non-accidental injury* or NAI, was the first form to be widely recognized (see *battered baby* syndrome) and is still the commonest form to be reported. However it is now recognized that other forms of child abuse may be at least as common, though often they are more difficult to identify. The major forms of abuse can be grouped under the headings of physical, emotional, and sexual, and in each case the abuse may be active or passive. See *failure to thrive, sexual abuse, neglect*.

Child-rearing styles A generalized term used to refer to characteristic ways of handling or dealing with one's children. The 1940s to 1960s saw a considerable amount of research into the effects of child-rearing or parenting styles, much of which proved inconclusive. One problem seems to have been that no account had been taken of the effects of the child on the parents. See *transaction*.

Chlorpromazine A widely used anti-psychotic drug, which is a derivative of phenothiazine. Chlorpromazine has a *sedative* effect, caused by raising the *threshold* for sensory information in the *brainstem*, reducing sensory input to the *reticular formation* of the brain, blocking the uptake of the *neurotransmitters* *adrenaline* and *noradrenaline* in the *sympathetic* section of the *ANS*, and also blocking the uptake of *acetylcholine* by *parasympathetic* nerve fibres. Chlorpromazine is sold under the trade names Largactil and Thorazine. See also *chemotherapy*.

Choice reaction time The time a subject takes to respond to a signal when the experimental conditions require a choice to be made. In general, reaction time increases as the number of choices increases in such a way that if the reaction time is plotted against the square root of the number or choices, a straight line will be obtained.

Chromatic colours Colours of varying wavelengths, which are perceived as having different hues (e.g. blue, red, yellow). See also *achromatic colours*.

Chromosome Strings of *DNA* which appear in the nucleus of a cell shortly before division as thread-like structures, arranged in pairs. Chromosomes carry the *genes* which determine the physical characteristics of the individual, and as such are large-scale units of heredity.

Chronic Continuing over a period of time. The term is usually applied to illnesses to distinguish persisting conditions (such as a depression which has been going on for years) from those that are expected not to last, or at least that have only just started and had a sudden onset. These conditions are called acute.

Chronological age The age of an individual measured by standard units, e.g. in months or years. In the original formulation of the *Intelligence Quotient*, by Binet, the measurement was obtained by comparing the chronological age of the child with its *mental age*. In this way, a comparison could be made as to how the child compared in learning skills with its contemporaries. (See also *mental age, IQ*). Chronological age is counted from birth and so may be misleading when applied to premature babies who are biologically younger than infants born at full term who have the same chronological age. The age counting from the date of conception is called the 'gestational age'.

Chunking The process by which, according to Miller, short-term memory can be extended. Miller's theory stated that short-term memory was of limited capacity, able to deal with only 7 plus-or-minus 2 items at a time. However, by grouping items of information into meaningful 'chunks' that capacity could be extended considerably (e.g. the figures 1, 0, 6, 6 would form four units treated separately, but just one 'chunk' if perceived as the date 1066). See also *short-term memory*.

Circadian rhythm A term used to describe bodily cycles that last for approximately 24 hours, e.g. of temperature and of alertness. Many individuals show pronounced circadian rhythms, becoming 'attuned' to their daily cycle. Disruption of such cycles, such as occurs when travelling from one time-zone to another, can produce an uncomfortable period of readjustment, known as *jet lag*. Extensive research by Kleitman and others has investigated natural human periodicity in cue-free environments such as caves in which lighting and temperature are kept constant. Physiological correlates of *diurnal rhythms* (e.g. fluctuations in body temperature) and the relationship between circadian rhythms and performance have been studied in this way. Circadian rhythms are also known as *diurnal rhythms* when referring to functions which occur during the day, and *nocturnal* rhythms for night-time activities. There is controversy over whether circadian rhythms are controlled by a *biological clock*.

Circular reactions Seen by Piaget (e.g. Piaget, 1959) as an essential mechanism of cognitive development during the sensori-motor stage. In circular reactions the result of an action triggers a repetition of that action, or some variation of it. As a result, actions are repeated and become practised and so competences are acquired and schemata are developed. At first they involve only the infant's own body and are called *primary circular reactions*. Later the child progresses to *secondary* and *tertiary circular reactions*.

Clairvoyance The perception of objects or events which are beyond the known reach of the senses. It is a particular form of *extrasensory perception*, distinguished by the fact that it is practiced by a 'medium', a person supposed to have special powers to communicate with and receive messages from distant or dead people. Clairvoyance is classified as a branch of *parapsychology*.

Classical concept A term referring to the classification of human concepts following work by J.S. Bruner and others on the development of thinking. Classical concepts are those in which the identifying properties of the concept are shown by every member of that class. So, for instance, all the cards of the suit 'diamonds' in a pack will show the diamond symbol, will be rectangular, etc. By contrast, although 'having four legs' would be an identifying property of the concept 'tables', not all members of the class would possess that identifying property. 'Tables' would therefore be a *probabilistic concept* rather than a classical concept.

Classical conditioning The procedure of pairing an originally neutral stimulus with a stimulus that reliably produces a response, so that the neutral stimulus comes to produce a version of the response on its own. In Pavlov's original experiment the neutral stimulus, called the *conditioned stimulus* or CS, was a bell which rang at the same time that the effective stimulus of food, called the *unconditioned stimulus* or UCS, was presented. Eventually the bell on its own came to produce some of the responses that food had elicited, such as salivation. These responses are called the

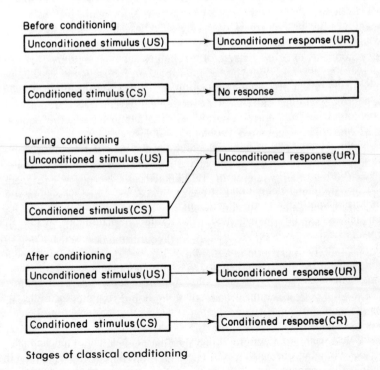

Before conditioning

Unconditioned stimulus (US) ⟶ Unconditioned response (UR)

Conditioned stimulus (CS) ⟶ No response

During conditioning

Unconditioned stimulus (US) ⟶ Unconditioned response (UR)

Conditioned stimulus (CS)

After conditioning

Unconditioned stimulus (US) ⟶ Unconditioned response (UR)

Conditioned stimulus (CS) ⟶ Conditioned response (CR)

Stages of classical conditioning

conditioned response or CR. The original full response to food was called the *unconditioned response* or UCR. Pairing an arbitrary CS with a UCR may require over 100 trials before conditioning becomes established. However when the UCR is a strong emotional response, such as fear, classical conditioning can be achieved in a single trial. Although it has been studied in the laboratory there are many everyday situations in which stimuli are paired in such a way that classical conditioning will occur. [f.]

Claustrophobia A fear of being in an enclosed or crowded area. But see *phobia*.

Client-centred therapy A form of *psychotherapy* developed by Carl Rogers, based on a *humanistic* approach, in which the individual is considered to be the only person who can develop solutions or approaches to their problem, and the role of the therapist is to facilitate such development. Because the therapist is frequently relating to highly approval-seeking individuals, the onus is strongly on the therapist to be *non-directive*, and to develop a genuine and warm relationship with the client, which will allow that individual to explore possibilities and options freely. See also *actualizing tendency*, need for *positive regard, unconditional positive regard*.

Clinical interview A method of investigation based on informal contact between the researcher and the individual(s) which he or she is studying. Use of the clinical interview technique avoids the main problem of artificiality in research, but sometimes at the cost of objectivity and *reliability*. It has been frequently used in psychology, for instance by Piaget in his studies of cognitive development in children.

Clinical psychology That branch of applied psychology which is concerned with the use of insights and methods obtained from theoretical psychology and clinical experience to assist those with problems in living, or with psychological difficulties. Over the last 25 years the profession has shifted from providing assessments as requested by psychiatrists to functioning as independent therapists. Clinical psychologists may use a range of techniques such as *cognitive therapy, behaviour therapy, psychotherapy, family therapy,* and *biofeedback.* The major specialisms are defined in terms of the client groups, i.e. general adult, child, mental handicap, neurology and the elderly. However clinical psychologists are increasingly to be found in community bases or working alongside general medical practitioners, and are beginning to be employed in industry.

Cloning A technique which makes use of the fact that the genetic 'blueprint' for a whole animal is reproduced in the genes and chromosomes of each cell nucleus in its body. By culturing small groups of undifferentiated cells, it is possible for them to develop into a complete individual of the same species, which is genetically identical to its parent animal. Successful cloning has been achieved on many different species of animal, ranging from frogs to sheep. The cloning of human beings to create a tightly stratified society forms a favourite theme of science fiction writers, but is unlikely to catch on in a big way, as the production of new human beings by traditional methods would appear to be both popular and effective.

CNS See central nervous system.

Coaction A term used to describe the process of acting jointly with another person, (e.g. working together on a task).

Cocaine A drug obtained originally from the coca plant, and used as a local anaesthetic. Freud is credited with reporting the first medical use. The drug also produces a sense of euphoria if taken internally and is often used as a recreational drug. It can produce *dependence*.

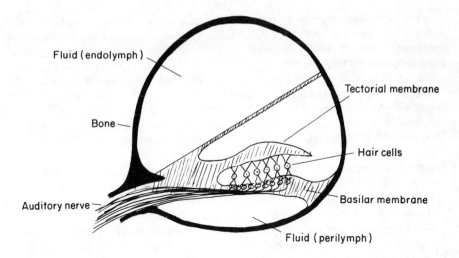

Fluid (endolymph)

Tectorial membrane

Bone

Hair cells

Auditory nerve

Basilar membrane

Fluid (perilymph)

Cross-section through the cochlea

Cochlea The coiled tube in the *inner ear* which contains the *organ of corti*. [f.]

Cocktail party problem A term given to one of the well-established phenomena of selective attention: the way that individuals are able to monitor unattended information subconsciously, such that they pick up highly sensitive information (e.g. their own name) even when attending to entirely different stimuli, and when they are unaware of the rest of the unattended message. See also *selective attention*.

Codes of language A description of styles of language use, which distinguishes two main 'codes' of language: *elaborated codes*, which involve a wide vocabulary and extensive use of nouns and explicit descriptions, and *restricted codes*, involving a more restricted vocabulary, a preference for pronouns, and the use of implicit description in preference to explicit. These codes were first described by Bernstein, who argued: (1) that elaborated codes were used far more by middle-class than by working-class individuals; and (2) that the language code used would facilitate or inhibit cognitive development, owing to elaborated codes being less dependent on context and therefore more amenable to abstract conceptualization. Bernstein's work was heavily criticized, notably by Labov.

Coding Also referred to as *encoding*, the term is generally taken to refer to ways in which information is represented cognitively, e.g. for storing in memory or for association with other information. Memories may be coded in a variety of ways, using many different modalities (e.g. *kinaesthetic*, or *enactive* coding, visual or *iconic* coding, auditory coding). See also *representation, schema*.

Coefficient A numerical value which indicates the strength of relationship, as in *correlation coefficient*. More generally a coefficient indicates how much a variable is modified, so a ball with a coefficient of elasticity of 0.9 keeps 90% of its momentum when it hits a hard surface and so will bounce more than one with a coefficient of 0.4

Coevolution A concept to recognise that natural selection is not a one-way process. While a species is adapting to its environment, the presence of that species will be having effects on the environment including other plants and animals. So *evolution* needs to be seen as a progressive mutual accommodation between species. The concept is similar to that of *transaction* in development.

Cognition A general term used to refer to the 'higher' mental processes. Cognition would generally be taken to include such kinds of mental activity as thinking and conceptualization; *memory, representation*, and mental *imagery; perception* and *attention*; reasoning and decision-making.

Cognitive behaviour therapy A method of psychological therapy derived from *behaviour therapy* but extended to take account of the patient's cognitions. The objective is to modify both maladaptive behaviours and maladaptive beliefs. See also *cognitive therapy*.

Cognitive development The way that *cognitions* develop during childhood. The major and most detailed theory of cognitive development is that produced by Piaget, though his theory is largely restricted to the ways thinking and understanding change through childhood. One of Piaget's most important contributions was to establish that the thought and logic of young children is not an inferior version of adult thinking, but has its own rules and is well adapted to the needs of the child. Cognitive development is not just a process of getting better at adult modes of cognition, but is a complex progression through different kinds of thinking and understanding. Other approaches to cognitive development include research into *metacognition, social cognition*, and the child's *theory of mind*.

Cognitive dissonance A concept put forward by Festinger, in which the main proposal is that each individual strives to maintain consistency between their differing cognitions. Should a noticeable inconsistency arise, this will produce a state of cognitive dissonance, which the individual experiences as uncomfortable and attempts to correct. Dissonance is reduced by adjusting one of the beliefs or attitudes involved in the inconsistency, so that the conflict disappears.

Cognitive map An internal representation of a specific or general area, which forms a plan or outline that can guide behaviour. The idea of cognitive maps was put forward by Tolman following work in which, for example, he demonstrated that rats which had been allowed to explore mazes freely would perform better when subsequently reinforced, than ones which had not had such an experience. Tolman used the concept of cognitive maps to illustrate one of the ways in which cognition might be involved in learning, at a time when learning was largely conceptualized as a reflexive, *stimulus-response* process. Later research on cognitive maps in humans demonstrated, for instance, the way that areas familiar to an individual would be perceived as larger and more complex than distant ones. Some cognitive theorists, among them Tolman, have argued that cognitive mapping forms the basis of all internal representation.

Cognitive psychology The branch of psychology which is concerned with the study of cognition. Cognitive psychology is generally taken to include the study of perceptual processes, *attention, memory, imagery, language, concept formation*, problem solving, *creativity*, reasoning, decision making, cognitive development and *cognitive styles*, but has often been assumed to exclude learning.

Cognitive styles Distinctive patterns of cognition which characterize individuals. Work on cognitive styles has included investigations of *convergent* and *divergent thinking, field dependence*, and forms of *intelligence*.

Cognitive therapy In its narrow sense, an approach to the treatment of *depression* developed by Aaron Beck. Beck sees depression as resulting from a combination of a negative evaluation of the self, a negative view of present experiences and events, and negative expectations of the future. The sufferer then uses faulty logic to maintain this outlook. The therapist must be very active to modify the way the patient thinks, insisting on correct logic and challenging unrealistically pessimistic assumptions. Beck has described specific techniques to be used in cognitive therapy but the term is now beginning to be used for a wider range of less well-defined approaches based on similar principles and applicable to a wider range of conditions.

Collective unconscious The concept, proposed by Carl Jung (e.g. Jung, 1964), that the human race has developed a shared unconscious mind which contains universal images called archetypes.

Colour blindness The inability to detect certain wavelengths of light. Most colour-blind individuals are red/green colour blind; that is, they are unable to distinguish between particular shades of red and their equivalent shade of green; but occasional individuals are blue/yellow colour blind. Colour blindness is found in about one in 10 males, although it is much rarer in females. It arises from a faulty gene carried on the X chromosome, which in women is normally counteracted by the normal equivalent gene on the other *X chromosome*. Males, however, have only one X chromosome, and the *Y chromosome* is shorter, so there is no chance of a 'healthy' gene to correct the fault.

Colour constancy The process by which the perceptual system compensates for the appearance of objects seen under light of differing wavelengths. Colour is detected by the analysis of the wavelength of the light reflected from an object. In normal white light, the light reflected will show the true colour, but under coloured lights an object may reflect light of a very different hue, owing to the mixture of colours. The brain, however, compensates for this by using its prior knowledge of the object and by *adaptation* to the viewing conditions, so the object is perceived as keeping its true colour.

Colour vision The ability to detect the specific wavelengths of light reaching the eye, which facilitates fine discrimination of detail and the use of colour as a signalling medium. Colour is detected to some extent by the cone cells of the eye, but the full mechanisms by which human beings detect colour are complex and as yet not fully understood.

Communication The process of transmitting information to another individual or group of individuals, and having it received and interpreted by them. Communication may be voluntary or involuntary; the individual who unwittingly signals that she is nervous by fidgeting, etc. is communicating that to the observer, although not voluntarily. Communication in human beings is complex and varied, and can be roughly classified into three general types: (1) *verbal communication* (using language or codes which stand for language); (2) personal *non-verbal communication* (such as the use of dress, posture, gesture or gaze to communicate); and (3) *ritual* (the use of highly structured events to communicate).

Community psychology The application of psychology to improving life for members of the community. The focus of community psychologists has been particularly on people whose capacity is reduced in some way, e.g. those who have lived for a long time in institutions. The term is used particularly for setting up environmental conditions such as sheltered housing which will make it possible for such people to live at least partly in the community.

Comparative psychology The branch of psychology which involves drawing comparisons between different species to give insight into the mechanisms of behaviour. Some psychologists see the value of comparative psychology as being to shed light on human functioning while others regard understanding animal behaviour as a legitimate goal in itself. Much of what has been called comparative psychology has in fact been the study of a single species of artificially bred laboratory rat. Comparative psychology includes many branches of learning theory (especially those in the *behaviourist* tradition), *ethology*, and any area of psychology which has been influenced by studies of animals (e.g. early theories of *attachment*).

Compensation Using other resources to make up for a deficit, as when a blind person makes exceptional use of sound stimuli. In psychoanalytic terms it is a way of overcoming, or at least concealing, a defect in personality, particularly in Adler's theory of compensation for feelings of inferiority. Note that compensating does not necessarily mean overcompensating. In neurophysiology compensation refers to the process in which an intact part of the brain may take over the functions of a damaged part.

Complex (i) A description implying that the phenomenon in question is complicated, probably having many influencing factors.

(ii) A noun used to describe a complicated mass, e.g. 'a complex of reasons'.

(iii) In *psychoanalytic* terms, a noun used to describe a set of emotionally charged phenomena and feelings, e.g. the *Oedipus complex*.

Complex cell A type of *neurone* found in the *lateral geniculate nuclei* of the *thalamus*, and in the *visual cortex* of the brain. Discovered by Hubel and Wiesel in 1968, complex cells form part of a hierarchical arrangement of cells which serve the function of coding incoming visual information into simple shapes and patterns. For a full description, see *simple cells*.

Compliance Conforming to accepted patterns of behaviour, or aquiescing in decisions. Kelman draws a distinction between conformity to others or to social norms arising from compliance, and conformity arising from the internalization of the group norms or values. Compliance is perceived as an outward conformity, with the individual reserving opinion or inwardly disagreeing. See also *conformity*.

Compulsion A repetitive, stereotyped behaviour which is both unnecessary and unwanted but which the individual still feels they have to carry out. Usually associated with obsessions, see *obsessive-compulsive disorder*.

Compulsive personality disorder See *obsessive-compulsive disorder*.

Computational theory of perception A theory developed by Marr (1982), who proposed that we identify objects as such because of various computations or calculations, applied by the brain to the visual image received by the retina. Computational theory emphasizes the characteristics of edges and boundaries in the visual image, and suggests that the brain uses these to build up an increasingly complex series of representations of the object until eventually a three-dimensional picture can be produced.

Computer simulation The use of computers to replicate human thought strategies and patterns of behaviours. Research on computer simulation has involved the study of the use of *heuristics* in reasoning, and of probabilistic judgements in decision-making. It is hoped by those involved that such research will eventually throw light on human *cognitive processes*. In *industrial psychology* computer simulation often provides a safer, cheaper, or more ethical way of examining what will happen to the process being simulated, under a variety of conditions. See also *artificial intelligence*.

Conative To do with intentionality. The conative domain was one of the three domains of the human psyche outlined by Galen, in the second century BC; the other two being the *affective* domain and the *cognitive* domain. This distinction has been maintained in attitude theory, where a given attitude is considered to have three components: an affective, or emotional component; a cognitive, or rationalized component; and a conative or behavioural component, which is concerned with the individual's tendency to act on the attitude in question. Conative means to do with will and intention, and in many ways represents a seriously neglected area of human psychology.

Concept A set of ideas and properties which can be used to group things together. It is a generalized idea which may be abstract (e.g. 'justice') or concrete (e.g. 'furniture'). Human cognitive processes are often considered to progress by the formation and elaboration of concepts, resulting from increased experience. See also *construct, classical concept, probabilistic concept, schema.*

Concept formation The name given to the process by which an individual comes to develop mental categories which will allow objects and events to be classified and grouped together. A considerable amount of research on cognitive development has emphasized concept formation.

Concrete operational stage This is the third of Piaget's four stages of *cognitive development*, characterized by the child's fascination with the material world and his strong inclination to collect facts and statistics. Children in the concrete operational stage were considered unable to deal fully with abstract concepts, and able to deal with those aspects of experience which had a material equivalent or which could be represented in a concrete fashion. The stage was considered to last from approximately 7 to 11 years of age. See also *sensori-motor stage, pre-operational stage, formal operational stage.*

Conditional positive regard A concept introduced by Carl Rogers, which refers to the satisfaction of the basic need for positive regard in human beings. The term 'conditional positive regard' refers to approval, love or respect given only as a result of the individual behaving in 'appropriate', or socially acceptable ways. A person who has encountered nothing but conditional positive regard throughout their life will, according to Rogers, become unable to satisfy the need for *self-actualization.* Autonomous action, or exploration of their own potential, necessitates taking a certain amount of risk, in that it could conceivably result in social disapproval. The formation of a relationship which provides *unconditional positive regard* for the individual provides the security for such self-realization to take place, and this is the goal of Rogerian *client-centred therapy.*

Conditioned reflex A physiological *reflex*, or automatic response, which is produced in response to a stimulus which would not normally produce such a reaction, but has come to do so as a result of the process of *classical conditioning.*

Conditioned reinforcer An event or stimulus which has acquired the property of strengthening a learned (conditioned) response, such that the learning is less likely to become extinguished. See also *secondary reinforcement.*

Conditioned response A response which is produced in specific conditions, as a result of being associated, through a training process, with a particular stimulus, known as a conditioned stimulus. The training process consists of repeatedly pairing a novel stimulus with one which will elicit the desired response automatically. After a while, the new stimulus will come to elicit the response independently, at which point the response is said to have become a conditioned response. See *classical conditioning.*

Conditioned stimulus A stimulus which brings about a response as a result of repeated association with an *unconditioned stimulus.* See also *classical conditioning, conditioned response.*

Conditioning A term used to describe the process of learning. Learning is considered by behaviourist psychologists to be the process of acquiring and reproducing specific behavioural responses under specific conditions: hence the term 'conditioning'. It is normally considered that there are two major forms of conditioning, usually referred to as *classical conditioning* and *operant conditioning.*

Conditions of worth A concept put forward by Carl Rogers concerning the way in which the individual's self-concept is affected by the *conditional positive regard* which he or she has experienced throughout life. Conditions of worth are an internalized set of values by which the individual assesses their own behaviour. In individuals who have experienced only *conditional positive regard* throughout life, such conditions of worth may come to represent unrealistically high standards of conduct, giving the individual a negative self-concept, and inhibiting the expression of their need for *self-actualization.*

Conduct disorders A group of *behaviour disorders* in children in which aggression or the breaking of rules is involved.

Cone cells Cone-shaped cells found in the retina of the eye which effect the *transduction* of light waves into electrical impulses which are subsequently transmitted to the brain for interpretation. Cone cells are concentrated particularly in the *fovea,* consequently, colour perception is better in the centre of the visual field. They contain a photosensitive pigment known as iodopsin, which breaks down on exposure to light. Different cone cells are maximally sensitive to light of different wavelengths. The two major theories of colour vision, the *trichromatic theory* and the *opponent process theory,* are both based on the fact that there are three types of cone, sensitive to different wavelengths of light and therefore responsive to three different colours, but the theories disagree about how colours are combined.

Conflict The result of opposed motives applying simultaneously. Most conflicts, for example between the desire to stay and finish an essay versus the duty of going out with friends, are easy to resolve. Some are much more difficult and result in an inability to act and the abandoning of both objectives. (If you really could not decide whether to write or go out, you might solve the problem by sitting in front of a piece of paper and daydreaming about going out). Difficult conflicts of various kinds have been studied experimentally, often with rats. *Approach–avoidance* conflicts in which a goal is both desired and feared are the most common, and readily result in inaction. Avoidance–avoidance conflicts (choosing between the frying pan and the fire) can easily occur and are very stressful if a choice has to be made. Usually of less concern are approach–approach conflicts, when going towards one desired goal means leaving another (the donkey that starved to death half-way between two piles of food).

Conformity The social process by which people in a group or in a social situation engage in behaviour which appears to be socially acceptable, that is, to go along with the social expectations apparent at the time. Conformity is often divided into *compliance* (conforming while inwardly disagreeing) and *internalization* (conforming as a result of internal agreement with the behaviour). Normative conformity refers to the process of conforming as a result of the existence of strong social norms directing the accepted behaviour; informational conformity is the process by which an individual may conform to others on the grounds that they are better informed about the situation; while ingratiational conformity refers to conformity with the specific purpose of the individual's achieving social approval, or a feeling of 'belonging'. The classic experiment in the field was conducted by Solomon Asch, who instructed groups of people to pretend to misjudge the length of a line and found that subjects in the group who had not received this instruction felt under strong pressure to conform. Conforming to group pressure is sometimes call the *Asch effect*.

Confounding variable A factor, or *variable* in a study which causes a change in the *dependent variable* (the measures being obtained); but which is not the *independent variable*, or main condition of the study. Many of the techniques of experimental methodology are ways of dealing with confounding variables. If, for example, the sex of the subjects is likely to influence a result, this may be dealt with: (1) by eliminating the factor (use only male subjects); (2) by controlling for sex (equal numbers in each group so the effect cancels out); or (3) incorporating it as a variable in the design (record male and female subjects separately and examine the effect of sex of subject on the dependent variable).

Congruence A general term used to refer to behaviour, attitudes or ideas which are in accord and not in conflict with other such behaviour, attitudes or ideas. See *balance theory*.

Connectionism The approach to *computer simulation* inherent in the use of *parallel distributed processing systems* to simulate human reasoning. The ability of such systems to produce novelty, in the form of unexpected or unanticipated outcomes, has been hailed as a major breakthrough in the development of *artificial intelligence*, although its value for adding to our understanding of human thought processes has yet to be demonstrated.

Connector neurone Neurones found in the *grey matter* of the brain and spinal cord which link and pass impulses on to other neurones. Connector neurones are also known as *relay neurones* or *multipolar neurones*. They are spidery in form, having

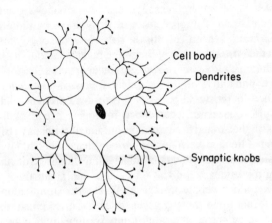

A connector neurone

several *dendrites* which enable the transmission or receipt of information to or from many other neurones. [f.]

Connotative meaning The meaning which is implicit in a particular term or phrase, although possibly not made explicit. See also *denotative meaning*.

Consanguinity Literally, being of the same blood, but also meaning that two people have genes in common, being descended from the same parent or member of a previous generation. The word usually appears in the context of marriage. Most societies forbid marriage between close relatives, as they are likely to produce genetic defects in any offspring. See *inbreeding*.

Conscience An acquired mental framework for making judgements about the moral 'rightness' or 'wrongness' of actions. The idea of conscience contains strong overtones of duty and obligation. It causes internal anxiety or distress to do things which conflict with the conscience. See *moral development, superego*.

Conscious That part of the mind which is readily available to awareness. Mental activities or contents that are not available are called *unconscious* when it is thought that they are being actively suppressed in ways originally described by Freud, and nonconscious when they are simply processes which are carried out without awareness. Processes that can be brought into consciousness but only with difficulty are called *preconscious* or subconscious.

Conscious process A mental process of which the individual is aware as it is happening.

Consciousness The awareness of one's own mental processes, or the state of having this awareness. The state of being aware of one's perceptions, thoughts, and feelings is vivid and undeniable but extremely difficult to study. The major issue is whether consciousness has any function or whether, as the behaviourists claim it is just a by-product of behaviour. As developments like *information theory* have provided a language for describing private mental events, psychologists are returning to the study of phenomena like consciousness. See also *unconscious*.

Consensus A common or generalized agreement, usually concerning *social norms* or acceptable behaviour; also used to refer to agreement between theories or ideas. In the *covariance* model of attribution theory, consensus is one of three factors believed by Kelley to influence the kind of *attribution* made for a specific situation or event. If a person is observed shouting, a relevant question is whether everybody else is shouting. If so, the condition of high consensus, it is assumed there is something about the situation that is responsible for the behaviour. If nobody else is shouting (low consensus) then the behaviour will tend to be attributed to some characteristic of the individual. See also *consistency, distinctiveness*.

Conservation The ability to recognize that volume, number or mass do not change when the physical appearance of the way that they are presented changes. In Piagetian theory, the ability to conserve is developed towards the end of the *pre-operational stage*. Prior to that time if the child is presented with, say, two identical balls of clay and one of them is rolled into a sausage shape, the child will say that the longer one contains more clay. Piaget considered this to arise from the process of *centration*: the child's tendency to focus on a single, central attribute of objects rather than taking several different aspects of its appearance into account. More recent studies (e.g. Donaldson), however, have demonstrated that the language used to the child, and the social situation of the experiments may have produced the result, and that children may be able to conserve at a much earlier age than Piaget suspected.

Consistency One of three factors in Kelley's (1973) *covariance* model of attributions. The more consistently a person produces the observed behaviour, the more likely we are to see it as arising from their disposition. See *consensus, distinctiveness*.

Conspecific Members of the same species.

Constancy See *perceptual constancy*.

Constancy scaling The process by which the perceptual system adjusts to distance, by mentally 'scaling up' objects which are far away, such that they are not perceived as being smaller. It is considered that constancy scaling may provide an explanation for certain visual illusions, e.g. the *Ponzo illusion*.

Constitution The inherited physiological structure of an individual. There have been attempts to relate psychological tendencies to the physical type of the individual, notably in Sheldon's theory of *personality*, but relationships are weak and inconsistent.

Construct A term used in *personal construct theory* to define *concepts* in a precise way. It is proposed that our cognitive system is made up of *bipolar* constructs such as illness–health and honest–dishonest. A large part of the theory is concerned with the relationships between constructs, e.g. a particular individual may have the idea that honest people tend also to be healthy. See *core construct*.

Constructive memory The general term given to memory for meaningful material which has been affected by the individual's own pre-existing *schemata, values* or *attitudes*. Since Bartlett, it has been observed that people rarely remember events or information accurately, but instead tend to adapt their memories to make more sense and accord with their own cognitions and cognitive styles, which is known as constructive memory.

Content analysis The *quantitative analysis* of verbal material, such as information obtained during interviews, from open-ended questions in questionnaires, or from reference material, like children's reading books. Content analysis involves firstly the identification of a number of defined categories – usually predefined by the researcher, but sometimes developed as a result of an initial inspection of the data. Once the categories have been defined, the number of times that each category of information occurs throughout the material is counted. In the case of more active data, such as the content analysis of video recordings, simple counting may be replaced by timing the duration of the behaviour or activity of interest. See also *qualitative analysis*.

Context The general setting or environment in which an event or a phenomenon occurs. There is evidence to suggest that *memory* is highly context-dependent, and that re-establishing a context will provide cues which facilitate the retrieval of memories. Similarly, the context of a communication or an utterance may be an important influence on how it is understood. See *state-dependent learning*.

Context-bound Limited to one particular setting and not applicable to others. The phrase context-bound is particularly used to refer to Bernstein's descriptions of *restricted codes* of language. He argued that the restricted code preferred by working-class language users is closely tied to the specific situation in which the utterance is made, owing to its reliance on pronouns rather than nouns and on nuances of tone of voice. This, Bernstein argued, serves to inhibit abstract conceptualization in the restricted language code user.

Contingency Any case in which one event has a raised probability of following another. In such circumstances an observer is likely to assume that the first event caused the second. *Reinforcement schedules* are examples of contingencies that have been experimentally manipulated. Research with infants has shown a high degree of alertness to events that are contingent on an action of the infant. For babies, contingent events are only likely to be provided by *caregivers* and so this alertness is believed to play an important part in orienting babies towards members of their own species. More generally, providing infants with contingent events has been suggested as a basic process of *attachment* and of the development of a sense of *self-efficacy*.

Continuity The expected consistency of various characteristics as the individual develops. Most developmental psychologists expected the *intelligence quotient* to stay reasonably constant as the child grew older, but it is now recognized that its continuity has been overestimated. In fact there is remarkably little continuity in any kind of measurable characteristics over anything more than short time periods. Most psychologists seem, like most other people, to believe in continuity and some are producing more sophisticated models of development to account for the lack of continuity in their data. See *transaction*.

Continuity hypothesis The belief that later functioning can be predicted from a knowledge of that individual earlier in their life. There are two forms of continuity hypothesis: (1) that characteristics such as personality and intelligence are relatively stable, so that if they are measured at one age, they will predict the strength of equivalent characteristics later in life; and (2) that significant events early in life will have permanent consequences, for example Freud's belief that early trauma is responsible for later psychopathology. This is known as the 'main effects' model; it can be contrasted with *transaction* as an alternative explanation for lasting effects. Continuity has been a major issue in *developmental psychology* because the evidence for either form of continuity has been very weak, despite good theoretical reasons and common-sense assumptions which suggest that it should be true.

Contralateral On the other side. Of interest to psychology because most of the brain's relationship with the rest of the body is contralateral, e.g. the left hemisphere controls the right hand.

Control group A group of subjects in an experiment which is used for comparison with an experimental group. The control group experiences all the conditions of the study in the same way as the experimental group, with the sole exception of the independent variable. In this way, by comparison of the results produced by the control group and the experimental group, the effects of the *independent variable* may be observed without contamination from the experimental situation itself.

Control processes Processes which use *feedback* in order to keep the functioning of a system within defined limits. The term has its clearest use in engineering and physiology and has been extended to psychological functions by analogy. See *homeostasis*.

Controllable attributions *Attributions* of a kind which imply that the person believes they have control or at least influence over the process. So, for example, believing that you passed an exam because of your hard work is a controllable attribution. In attributional analysis, it is not always clear whether the controllability is intended to apply just to the cause, just to the outcome, or both.

Conventional morality This is the second of the three stages of *moral development* proposed by Kohlberg. Individuals at this stage consider that society's rules are by definition moral. In the early part of the stage, the individual adopts moral codes in order to avoid social sanctions. In the second part of the stage, such moral codes or rules are seen as intrinsically right because they facilitate the smooth operation of society, and therefore should not be challenged. See also *autonomous morality, pre-moral stage*.

Convergent thinking Problem solving which works consistently towards a defined solution; a way of thinking that assumes there is a single right answer and that the way to reach that answer is to work directly towards it. It has been pointed out that within the educational system students are trained in convergent thinking and that *intelligence tests* depend entirely on convergent thinking ability. Rather less justifiably it is then assumed that convergent thinking is opposed to *creativity* and is inferior to creative, or *divergent thinking*. It could be argued that the reason that most people use convergent thinking most of the time is because it works for most problems.

Coping behaviour A general term given to behavioural strategies or techniques which allow an animal or human to reduce the amount of stress experienced in a given situation.

Core constructs A term used in *personal construct* theory to describe those *constructs* which are most closely associated with a person's self-concept. Core constructs are ones with which the individual identifies strongly and which tend to be utilized in a wide variety of situations.

Corpus callosum The band of neural fibres which connects the two hemispheres of the cerebrum. *Split-brain studies* involve the study of behavioural and learning changes produced when the corpus callosum is cut, such that the two hemispheres operate independently and cannot pass information to each other.

Correlation A statement of a relationship between two *variables*, such that changes in one tend to be accompanied by changes in the other. In a positive correlation, when one variable increases the other increases, e.g. tall people tend to be heavier while shorter people tend to be lighter. There is therefore a positive correlation between height and weight. If two variables show a negative correlation, then as one increases the other decreases, e.g. reaction times get longer as the visibility of the stimulus diminishes. A perfect positive or negative correlation will show changes in the two variables which are exactly proportional to one another; a weaker correlation will show more variability in the extent to which the two measurements match up. See also *correlation coefficient, scattergram*.

Correlation coefficient This is a numerical statement of the extent to which two variables vary together. A correlation coefficient is expressed as a number between +1 and −1, with +1 representing a perfect positive correlation (i.e. when one variable increases, the other shows an increase which is precisely proportional to it), and −1 representing a perfect negative correlation (i.e. one where a decrease in one variable shows a precisely proportional increase in the second). In situations where there is little or no relationship between the two measurements, the correlation coefficient will be close to zero. See *Pearson's product–moment correlation; Spearman's rank–order correlation*.

Cortex A general term used to refer to the *cerebral cortex*.

Cot death See *sudden infant death syndrome*.

Counselling The term has two rather opposed meanings. (i) Counselling is a form of therapy derived from the *non-directive counselling* of Carl Rogers in which the client is supported while they gain insight into their problem and work on finding their own solution. Within this use, people who offer therapy but who have no formal qualification or whose therapy is carried out as part of another job (e.g. priests), usually call themselves counsellors.

(ii) Counselling is also guidance on practical personal problems such as vocational choices, problems in studying etc. These counsellors are much more active in providing information, offering advice etc. In North America this kind of work is an established career and the practitioners are called counselling psychologists.

Counterbalancing A strategy used in the design of those experiments in which it is possible that the order of presentation of the conditions of the study could produce an unwanted effect, such as a *practice effect* or a *fatigue effect*. Counterbalancing involves systematically varying the order of presentation of the conditions, such that, for example, half of the group of subjects would have condition A followed by condition B, while the other half would have condition B first, followed by condition A. See also *order effects*.

Counter-conditioning In *behaviour therapy*, the *conditioning* of a response which is incompatible with an existing undesirable behaviour. Someone who is afraid of spiders might be trained to relax whenever they think of a spider, so their first reaction prevents them from feeling fear.

Counter-transference In psychoanalytic therapy, but presumably occuring in many other contexts as well, the feelings produced in the therapist by the patient. Regarded as a valuable clue to help understand what is happening to the patient and the effect they have on people in their outside relationships: if the therapist notices feelings of irritation or a wish to protect the patient this can be used to help the patient understand what is going on in the session and to clarify the effects they have on other people; it will also help the therapist identify the nature of the patient's *transference*.

Covariance (i) In *attribution* theory covariance is a central concept in Kelley's approach. It consists of a way of predicting what attribution a person will make for events by taking account of three kinds of information that they may use: *consensus, consistency*, and *distinctiveness*. The theory is also known as the *ANOVA model of attributions*.

(ii) In *statistics*, covariance refers to the situation in which a change in one variable is accompanied by a change in the other. The term avoids specifying which variable affects the other, and leaves open the possibility that there is no direct causal connection at all. For example the number of blankets sold in Canada covaries with the number of people having colds in England. This is not because colds are caused by other people's blanket-buying behaviour, but because a third factor is affecting both phenomena. If one variable is regarded as an unwanted influence on the other, statistical techniques can be used, which adjust the information in such a way as to counteract its effects, at least in theory.

CR See *conditioned response*.

Creativity The ability to produce novel products or solutions to problems. Creativity has been studied as a counterpart to intelligence, represented by *divergent* and *convergent* thinking abilities respectively. However, it is difficult to devise tests, as a creative response is by definition unpredictable, so correct answers cannot be specified in advance. In fact there is no agreed way of measuring how creative any particular achievement may be. Also it is probably even less appropriate than with intelligence to think of creativity as a quality of which an individual has a certain measurable amount. Despite these difficulties E. Paul Torrance has produced a test of creativity that seems to work quite well (it includes classic items like 'how many uses can you think of for a brick'). He claims that results from the test show that school education reduces the child's creativity. The classic theory of creativity is that it requires preparation (doing the ground-work), *incubation* (a period of sub-conscious processing), inspiration (a sudden insight), and verification (checking the solution works). More recent theories, for example Edward de Bono's, usually come down to claiming that creativity results from a random element in thinking. It seems unlikely that Leonardo da Vinci's output could be accounted for in this way. So the present state of the field is that we have no plausible theory of how creativity happens, no reliable way of measuring the creativity of a person, and no real idea whether creativity happens because of characteristics of the individual or because of particular kinds of circumstances. We clearly need a creative solution to these problems, but we do not have much idea of how to achieve this.

Cretinism A severe *congenital* condition caused by a lack of thyroxin, sometimes because of a lack of iodine in the pregnant mother's diet. The result is severely stunted physique and brain development.

Crib death See *sudden infant death syndrome*.

Criterion A standard or yardstick by which a judgement or evaluation is made. One use of the term is for the level of probability required for a statistical result to be regarded as significant. The usual criterion is a probability level of less than 0.05.

Critical period A time period during the development of the individual in which a particular function can readily be acquired. Outside of the specific age-range it will be difficult or impossible to acquire the function. The function may result from physical development (*maturation*) or from *prepared learning*. *Imprinting* in ducklings is a well known example, and in human infants if three-dimensional vision is not achieved by the age of about two years then it will never be acquired. On a strict definition, a critical period should be a well defined time during development and the function should be impossible to achieve either before or after this period. However, outside of physical growth processes, examples of strict critical periods are rather rare. It is now known that even imprinting can be obtained well after the end of the normal critical period. In human development it is now more common to speak of sensitive periods, but even this looser term has often been applied too enthusiastically. For example, it is not very helpful to refer to a critical or *sensitive period* for language acquisition when language can be acquired at any time during a period of at least 12 years and possibly more.

Cross-cueing The process observed in *split-brain* patients by which one hemisphere of the brain transmits information to the other. In a typical experiment, a subject may be shown an object to one side of the brain only. Later, the object is shown to the other side of the brain, and the subject is asked questions about it. Although in such patients the *corpus callosum* has been cut so no direct transmission of information between the cerebral hemispheres is possible, subjects may produce feedback on the correctness of the answer offered by an imperceptible nod, frown, or other physical signal. This is recognized by the other side of the brain, so that the question can be answered correctly. Cross-cueing of this nature can often be extremely rapid and subtle.

Cross-cultural study A study which involves the comparison of people from different cultures.

Cross-modal transfer The transferring of information from one sensory mode to another. For instance, figure-ground perception learned as a result of experience with touch may also be applied when the subject is using vision. This kind of transfer is found frequently with subjects who have acquired a new sensory function; e.g. people blind from birth who have obtained their sight through an operation performed in adulthood.

Cross-sectional study A method of research in developmental psychology which involves comparing individuals of different age groups, e.g. measuring the moral judgements of six year olds, ten year olds, and fourteen year olds. Such an approach is cheaper and easier to carry out than a *longitudinal study* in which the same children would be repeatedly measured at different ages, but may present other problems. A further example is the study of psychological decline with ageing. Older people may be found to have lower *intelligence quotients* than a younger sample, but this can be because their diet and education as children in 1920 were inferior to the diet and education available to the younger sample who were children in 1970. So age differences found through cross-sectional samples may not be a direct result of the ageing process.

CS See *conditioned stimulus*.

Cue Something which gives an idea or a hint about something. More specifically, a cue is information which activates a *schema*. A cue in *memory* theory, for instance, is a remembered item which connects with further information, allowing the individual to retrieve more. In *perception*, a cue is the item of information which is used by the brain to direct the interpretation of specific stimuli: a depth cue is that part of the information which indicates how far away something is.

Culture A general term used to describe the set of accepted ideas, practices, values and characteristics which develop within a particular society or people. Although most modern societies are *multicultural* to some degree, the word 'culture' is often, though not accurately, used interchangeably with 'society'.

Culture-free and culture-fair tests During the 1960s and early 1970s considerable efforts were made to develop *psychometric tests*, e.g. *IQ* and *personality tests*, which would avoid cultural bias by being free from reference to culture altogether. In practice the diversity of cultures was so great that such tests proved impossible to develop. Researchers had to content themselves with the attempt to establish tests which, instead of being completely free of cultural influences, allowed a fair assessment of those from other cultures. Such culture-fair tests are psychometric tests which do not provide an advantage to members of one culture over another. In practice, however, culture-fair tests are extremely difficult to achieve, owing to cultural diversity which not only produces differences in background knowledge and skills, but also in motivation and attitudes to tests. It is very difficult for those compiling the tests to be fully aware of their own cultural assumptions. It could also be argued that since the culture itself is not fair, a biased test will give more accurate predictions: a test which gives an advantage to middle-class academic values will more accurately predict which children will do best in school.

Culture-specific Occurring in, or belonging to, a particular culture.

Cumulative record or cumulative curve A graph in which each successive point shows the total number of responses up to that time. It is mostly used to show the progress of *operant conditioning* and has the advantage that the steepness of the curve gives a direct indication of the rate of responding.

Curare A paralysing poison used in blowpipes by some South American Indians, for hunting. Curare achieves its effect by being picked up at *receptor sites* in muscle fibres, thus blocking the uptake of acetylcholine such that messages from the central nervous system are not received. Curare therefore prevents voluntary muscle action but does not affect the actions of involuntary muscles. Animals which have been killed with curare die from suffocation, but if respirated artificially until the curare has worn off, will stay alive. Consequently, curare has proved useful in several psychological studies investigating, for example, the effects of muscle actions on cognitions.

Cutaneous To do with the skin.

D

Dark adaptation The process by which light-sensitive cells in the *retina* come to adjust their sensitivity to light, such that they will fire even in response to very faint stimuli. Dark adaptation in the human being takes approximately 20 minutes, beginning with a rapid period of adaptation while the cone cells adjust, followed by a longer period for rod cells to obtain maximal sensitivity.

Data A general term for all forms of recorded information. Usually the term is used for the scores obtained in a survey or an experiment. Note that 'data' is a plural word, the term for a single score being 'datum', so it is wrong to write 'the data is . . .'. Unfortunately there is no standard singular word for a collection of data, but 'result' or 'information' will often do.

Daydreaming The activity of engaging in fantasies or imaginative speculations during quiescent waking periods. Some research suggests that daydreaming may be instrumental in promoting positive mental health for the individual, perhaps through the clarification of goals and ambitions.

DB See *decibel*.

Decentration The process by which an individual is able to step out of their own mental perspective, and to take another person's point of view. According to Piaget, the ability to 'decentre' only emerges during the *pre-operational stage*, and forms a part of the gradual reduction of *egocentrism* which Piaget saw as central to cognitive development.

Decibel (DB) A unit used to express the loudness of a sound. The decibel scale involves a progression which is nearly logarithmic, such that doubling in perceived intensity is represented by an increase of approximately three units on the decibel scale.

Decision theory Any theory which attempts to explain how decisions are made. In practice the term is most often applied to theories which apply mathematical models to human decision processes. See *receiver-operating-characteristic curve*.

Deconstructionism The process of identifying how scientific theories, not only in psychology, come to reflect the social assumptions and conventions of their time, or of those propounding the theories. So, for example, in deconstructionist terms, the association of Lorenz with the German Nazi party would not be seen as unconnected with the theory of aggression which he propounded.

Deep structure A term coined by the linguist Noam Chomsky to describe the universal properties of basic grammar, supposedly common to all languages. It was the similarities of deep structure which allowed for Chomsky's proposed innate *language acquisition device*, a theoretical construction by which he explained the infant's readiness to acquire human language.

Defence mechanism A strategy which protects the ego or self-concept from real or imaginary threat. First proposed by Freud, defence mechanisms may take many forms, of which a few of the most common are: *repression, reaction formation, projection, rationalization, identification* with the aggressor, and *denial*. Although Freud classified defence mechanisms as *neurotic* or *psychotic*, the fact is that everybody uses them sometimes as a way of avoiding unwanted information about themselves or the outside world. They all have the disadvantage of distorting one's understanding of reality.

Deficiency motive A motivation that arises because of a perceived deficiency of some kind. The deficiency can range from physiological (e.g. food) to higher needs, such as that for recognition. Deficiency motives are distinguished from 'abundancy motives' in which it is judged that the organism is trying to acquire more of the material than is needed for comfortable survival.

Degeneration In neurophysiology, the deterioration of neural tissue which occurs through lack of stimulation, injury, or lack of nutrients. In *stimulus deprivation* studies, some damaged perceptual functioning which was originally thought to result from cognitive deficits was later found to be caused by neural degeneration.

Deindividuation The process by which individuals come to feel that they are simply part of a corporate entity, such as group or crowd members. Deindividuation involves the individual's surrendering the immediate perception of independence and autonomy, and feeling as though they have merged anonymously with the other people involved. It is commonly found in military units in action, and in mobs. See *diffusion of responsibility*.

Delayed conditioning A form of *classical conditioning* in which the conditioned stimulus is presented several seconds before the unconditioned stimulus, but with both coming to an end at the same time. By comparison with *simultaneous conditioning* or *trace conditioning*, delayed conditioning is considered to be the most effective.

Delta waves Long slow wave patterns which can be observed on the *electro-encephalograms* of people in deep sleep. Delta waves begin to appear during the third level of sleep, and are most common during level IV sleep. See also *sleep cycles*.

Delusion A belief which is mistaken, yet is held firmly despite contradiction by evidence and logic. Delusions of grandeur may occur in schizophrenia, and delusions of persecution are common in paranoia.

Demand characteristics Those features of an experimental or similar setting which elicit unusual or situation-dependent forms of behaviour from subjects participating in the study. These would include factors such as *experimenter bias*, expectations held by the subject as to the 'correct' way to behave when participating in a psychological study, the effects of trivialized or meaningless tasks which necessitate uncommon strategies to deal with them, and the wish to give the experimenter the result she wants.

Dementia A state in which the cognitive abilities of a person are so damaged that they are no longer able to function independently. The term is now used almost exclusively for permanent physical deterioration of the brain. The most common form is senile dementia, which occurs in old age but has a variety of causes. See also *Alzheimer's syndrome*.

Dendrites The branched structures at the end of the *axon* of a *neurone*, which are used for the transmission or reception of *neurotransmitters*, and so contribute to either the excitation or the inhibition of the *electrical impulses* through synaptic transmission. A dendrite will characteristically end in a swelling, or synaptic knob, which carries vesicles containing a neurotransmitter. Receptor sites on the dendrites of the adjoining neurone pick up the neurotransmitter.

Dendron That part of the elongated stem of a *neurone* found before the cell body, taking the same direction as that in which the impulse travels.

Denial A *defence mechanism* or aspect of one's own psychological functioning of not acknowledging the existence of a threatening event or utterance. Denial is most commonly found in children, although it is not uncommon as an adult defence mechanism.

Denotative meaning The specific or symbolic meaning of an utterance or term. The denotative meaning of something is that which is simply and necessarily contained in the use of that term, without any of the additional associations or implications which a listener may understand. See also *connotative meaning*.

Dependency (i) A term used to express an unbalanced relationship in which one individual relies consistently on the support or aid of another. In this sense the term was formerly used to describe the relationship of an infant to its mother, but has now largely been replaced by *attachment*.

(ii) A term used to describe a reliance on a particular drug or therapy, which falls short of physiological *addiction* but which is characterized by a psychological reliance such that the individual feels unhappy or uneasy in its absence.

Dependent variable The variable which is measured as an indicator of the outcome of an experiment. If you set up an experiment to assess the effect of coffee on speed of essay writing, the dependent variable would be the measure of writing speed. The dependent variable is so named because, if the experimental *hypothesis* is valid, its value will depend on the condition of the *independent variable* which has been set up.

Depressant A drug which reduces or depresses physiological functioning, particularly central nervous system activity. *Alcohol* is the most widely available depressant, though its effects may be concealed temporarily by its capacity to induce euphoria. The term may also be applied to psychological influences which have an effect of lowering mood. See also *stimulant*.

Depression A reduced state of both physiological and mental functioning, usually associated with feelings of unhappiness. The most common symptoms are a loss of interest and inability to enjoy any experiences, sadness, loss of appetite, sleep disturbances (especially early in the morning), passivity, and suicidal thoughts or intentions. However even very severe depressions may only involve a few of these symptoms. The term is used for a very wide range of conditions extending from 'ordinary unhappiness' through to psychotic disorders. Psychologists will therefore indicate when they are using the term to refer to a serious clinical condition either by the context or by attaching further labels: either 'clinical depression' or a specific term for a particular form of depression. The more common of these are: *bipolar depression, endogenous depression* (caused internally), exogenous or *reactive depression*, and *psychotic depression*. Other forms of depression are 'agitated depression' in which the individual is agitated, restless and irritable, and 'retarded depression' when they are slow, apathetic and difficult to get moving.

Depth cue A perceptual factor which gives an indication of how far away an object or image is. See *depth perception, monocular depth cues*.

Depth perception The interpretation of distance from sensory information. Depth perception relies on two main sets of depth cues: *binocular depth cues* and *monocular depth cues*. Binocular cues include *retinal disparity, convergence* of the eye muscles and accommodation of the lens, while monocular cues include height in plane, superposition, shadow, gradient of texture and colour, relative size, and motion parallax. Auditory depth perception involves the interpretation of attenuated signals, such that sounds which are further away are fainter; and also of phase shifts in the wavelengths of sound, such that sounds which come from further away appear to be muffled by comparison with nearer ones.

Desensitization A procedure which will reduce the responsiveness of the subject. Used mostly for behavioural techniques which reduce or eliminate inappropriate emotional responses, usually anxiety. The basic procedure is to present weak forms of the feared stimulus while using stronger forms of a stimulus that produces a response which is incompatible with anxiety. The feared stimulus is then gradually increased in strength without triggering the fear response. The standard procedure is called 'systematic desensitization' and is an example of *counter-conditioning*.

Deoxyribonucleic acid (DNA) The chemical which forms the basic units of chromosomes and is therefore fundamental to reproduction.

Development The processes of change over the life span. One aspect is physical development which is strongly influenced by genetic tendencies. The other is psychological development which is much more directly influenced by environmental factors.

Developmental disorders Disorders which appear to result from a failure of developmental processes and which can be expected to distort future development.

Developmental norms The expected level of performance of children at a specific age. For example in a given population the norms for the number of words spoken might be 50 at age 18 months, 400 at three years etc. Developmental norms can be used to give a precise indication of how uncommon any unusual performance by a child may be. Identifying a level of performance as being exceptionally poor is only the first step in deciding whether any further action is desirable.

Developmental psychology The psychological study of development. Some distinction is made between developmental psychology which is the study of the laws and processes of development, and child psychology which is more focused on empirical techniques for studying children at specific ages. However the terms are often used fairly interchangeably, and the phrase 'experimental child psychology' has come into use to preserve the distinction. Major theories of development have been propounded by Freud, Gesell and Piaget, among others. All of the large-scale theories were established in the first half of this century, and most are restricted to childhood. However there is reason to believe (or at least hope) that development continues throughout adulthood. The field of *life span* developmental psychology has therefore become active in recent years but as yet has no major theory as a basis. In fact developmental psychology in general seems to be proceeding quite adequately at present without much reliance on overall theories of development. Instead there are theories to deal with restricted areas such as *attachment* and *language*, and a focus on a number of more or less practical issues. The areas of greatest current interest include: the growth of cognitive and social competence; the *nature–nurture* or genetic–environment debate; the question of *continuity*; the way a child develops a *theory of mind;* applications to education and to parenting; the importance of *play* and *creativity*; and most recently, the family.

Dialect A distinctive pattern of grammatical forms and vocabulary which originates from a particular region. The point where a dialect becomes distinctive enough to be seen as a language in its own right is one of social judgement, rather than any linguistic criteria: some linguists for instance, regard the West Indian Creole dialect or Hong Kong English as distinctive languages in their own right, since although they may have originated as forms of English, they contain their own distinctive grammatical forms and vocabularies. The same situation pertains to a number of European languages, such as Flemish, where considerable social action was required in order for it to be regarded as a separate language rather than a regional dialect. It may be observed therefore that the social recognition of an extremely distinctive shared form of speech as a language rather than a dialect has everything to do with the acknowledged social status of the group which uses that form of language, and relatively little to do with the linguistic structure of the form of speech itself. See also *accent, speech register, psycholinguistics*.

Dichotic listening task A method for investigating selective attention by presenting two different messages through the two sides of headphones, and asking the subject to attend to one only. Dichotic listening tasks are usually monitored by asking the subject to engage in 'shadowing' – speaking the attended message out loud as they listen. [f.]

A dichotic listening task

Dichromatism A term used to describe forms of colour vision in which the individual is lacking in sensitivity to specific wavelengths of light. Normal colour vision is trichromatic, in that three major wavelengths go to make up any given colour, but some colour-blind individuals use dichromatic vision in interpreting specific hues, i.e. using two major wavelengths only. See *colour blindness*.

Difference threshold See *relative threshold*.

Diffusion of responsibility The process by which individuals may fail to act in a situation requiring *bystander intervention* as a direct result of the presence of several other onlookers. The perception is that this implies that the responsibility is shared, which reduces the pressure on each separate individual to act.

Digital Coded in simple on–off (binary) units, such as in computers and the traditional view of the firing of neurones.

Directed thinking Thinking which is directed towards a particular goal, e.g. in problem-solving.

Discourse analysis A general term covering various ways of analysing spoken or written communication. The term 'discourse' avoids the assumptions built into terms like conversation. There are a number of techniques of discourse analysis, such as identifying the recurrent *semantic* themes of a discourse, or the use of metaphor. Linguists apply a more specific meaning to the term, relating to natural breaks in the discourse. See also *content analysis, social representations*.

Discovery learning A form of educational practice studied particularly by J.S. Bruner, in which students operate mainly by deduction and inference, with guidance and resources being provided by the teacher. Discovery learning emphasizes the student's own activity and enquiry, rather than the teacher's transmission of information.

Discrimination (i) The skill of distinguishing one stimulus from another; usually learned through selective instrumental or classical conditioning.

(ii) The practice of drawing arbitrary distinctions between one set of people and another, such as is found in a group of highly prejudiced individuals taking steps to limit or restrict access to privileges or resources by a minority group.

Discriminatory stimulus A stimulus in operant conditioning which provides a cue to indicate when a particular response is appropriate or not.

Disembedded thought Thinking which is not applied in a relevant context, but is required to take place independent of context. Many of the criticisms put forward of Piagetian approaches to the understanding of the child's cognition centre around the idea that the child was required to engage in disembedded tasks; and that when the tasks were put in an appropriate social context, children were noticeably more successful at them. See also *naughty teddy*.

Disengagement A theory of ageing put forward by Cummings and Henry in 1961, in which it was proposed that the elderly undergo a process of systematic disengagement or withdrawal from society, reducing the amount of participation in and integration with society. The process was thought of as a way of coping with the deaths and illnesses of partners and friends, and as a possible preparation for approaching death. Cummings and Henry proposed that this behaviour had a possible biological origin. The theory has been heavily criticized, mainly on the grounds that social pressure on old people to withdraw from society is high, and that for many, society affords few alternatives to withdrawal. *Social exchange processes* have been suggested as possible alternative models.

Displaced aggression Aggressive behaviour which is directed towards a target which is not the original source of frustration. Typically, aggression becomes displaced because the original target is unreachable, or because it would be inexpedient for the individual to direct aggression towards the original. For instance, it may be dangerous for someone to express directly the aggressive feelings generated by an unpleasant boss, and such feelings may become displaced onto family members instead.

Displacement A process of channelling undesired or inexpedient impulses to alternative outlets. An example would be the application of aggressive tendencies to becoming the best chess player in the college. When the outcome of displacement is regarded as socially desirable the process is also called *sublimation*.

Disposition A tendency to behave in a particular way. When used by developmental or clinical psychologists the term implies an inherited tendency, and is used interchangeably with predisposition. When used in the context of motivation and personality it is a general term for any relatively stable behavioural tendency, and no genetic basis is implied.

Dispositional attribution Believing that a person's behaviour is caused by their character or personality, rather than the situation that they are in. People are usually more likely to make dispositional attributions about the behaviour of other people, and to account for their own behaviour in terms of the situation they were in. See also *attribution error, situational attribution.*

Dissociation A separation of two parts of an individual's mental life so that each can function separately or even in contradiction to the other. Extreme forms are *amnesia* and *multiple personality*, but milder forms are more common. For example, when someone is competitive at work but not at home.

Dissonance A state in which a cognitive discrepancy is produced between two events, such that one cognition is in direct contradiction to another. Typically, such *cognitive dissonance* results in attitude change, such that the dissonance is reduced.

Distance cues See *depth cues.*

Distinctiveness A concept in attribution theory which concerns how unique an event or behaviour is. Distinctiveness is one of three major criteria used to formulate attributions for any given situation. The other criteria are *consistency*, and *consensus*. If a person is shouting on a particular occasion we might ask whether they usually shout in other contexts as well. If not, the condition is one of high distinctiveness, and we tend to assume there is something about the situation which is producing the behaviour. If the person always shouts (low distinctiveness) then we attribute the behaviour as a characteristic of the person.

Distributed practice A procedure during learning in which time gaps are interspersed during the practice. For example, if you were trying to learn the contents of a chapter, you would take a short break at the end of each page. This approach has been found to lead to more effective learning than *massed practice* in which no breaks are taken.

Distribution The pattern made by a set of scores when grouped according to frequency. Theoretical distributions are the pattern that would be produced by scores that conformed precisely to a mathematically defined function. The most important of these is the *normal distribution*, but each statistic has its own distribution.

Diurnal rhythm A biological rhythm in which activity and alertness peak during the daytime. It is a form of *circadian rhythm.*

Divergent thinking Thought which ranges far more widely than is conventional. Tests of divergent thinking are often included in *creativity* tests, as it is assumed that highly creative individuals will be able to utilize novel frameworks more readily than those with a more conventional style of cognition. See *convergent thinking.*

Dizygotic twins Twins which have developed as a result of the simultaneous production of two ova by the mother, both of which have subsequently been fertilized and developed fully. Unlike *monozygotic* twins, they resemble each other genetically only to the extent that ordinary brothers and sisters do. Dizygotic twins are also known as fraternal twins.

DNA See *deoxyribonucleic acid.*

Dominance A term used loosely by ethologists to refer to privileged access to resources, rights of way, or appeasing treatment by other members of a social group.

Dominance hierarchy A concept first proposed in 1922 by Schjelderup-Ebbe, after observation of a consistent order of precedence (the pecking order) among hens when given restricted access to food supplies. Dominance hierarchies became popular as ethological concepts throughout the 1950s and 1960s, and were considered to present a basic model of social organization for most social animals, but the existence of linear dominance hierarchies has been increasingly called into question by ethologists in recent years.

Dominant gene A gene which is more likely to be expressed in the individual's development than a matching gene (*allele*) with a different physical implication. For instance, if an individual inherits a gene for red hair and a gene for dark hair from its parents, the dark haired gene, being dominant, will be the one that is expressed. Red hair genes are *recessive* and will only be expressed in the phenotype if both alleles code for red hair.

Dopamine A neurotransmitter involved in reward and motivational pathways in the brain, and possibly implicated in some psychiatric disturbances. The tranquillizer *chlorpromazine* (Largactil) seems to work by blocking dopamine receptor sites; while *amphetamines* effect an increase in the levels of dopamine and *noradrenaline*. The symptoms of *Parkinsonism* can be alleviated by the drug L-dopa, which increases dopamine levels in the brain. The implication here is that a dopamine deficiency may be causing the disease.

Dopamine hypothesis The hypothesis that *schizophrenia* is caused by an excess of dopamine in the limbic system.

Double bind A situation in which the individual appears to be confronted with alternatives but in fact whatever they do will be wrong. For example a father might forbid his son to climb a tree. If the boy climbs the tree he is punished for disobedience, but if he does not, his father indicates that he is disappointed at the boy being so 'soft'. Double binds seem to be particularly common in families, and Gregory Bateson, who invented the term, initially proposed that schizophrenia was caused by growing up in a family in which double binds were used frequently. This theory has now been abandoned but systemic *family therapists* recognize that double binds are a common and destructive feature of many disturbed families.

Double-blind control An experimental control in which neither the person conducting the experiment nor the subjects participating in the study are aware of the experimental hypothesis or conditions. Double-blind controls are necessary as precautions against *experimenter effects*, and are considered essential in tests of new drugs or assessments of therapeutic procedures.

Down's syndrome A congenital disorder in which the individual possesses an extra chromosome, giving rise to a series of distinctive physiological characteristics, often accompanied by *mental retardation* and language difficulty. Also called mongolism, a term which originated because of a slight resemblance of children with Down's syndrome to Mongolians, and is now no longer used in the psychological literature.

Dream Mental activity which occurs during *sleep*. Dreams typically have vivid imagery, an emotional content, and occur during a particular phase of sleep (*rapid eye movement sleep*). They also have the characteristic of being rapidly forgotten on waking. It seems that all humans dream but most dreams are not remembered. Freud proposed that the function of dreams was to preserve sleep by seeming to fulfil wishes that would otherwise disturb the sleeper. More recent theories propose that dreams are the by-product of the processing of information that has come in during the day and needs to be incorporated into the cognitive system. However some obvious predictions from these theories have been contradicted by research, and both the nature and function of dreams remain something of a mystery.

Dream analysis Finding hidden meanings in disguised symbolic form by interpreting the content of dreams. Dream analysis is a particular tool of the *psychoanalytic* schools of thought proposed by Freud and Jung. It is considered to form an important set of clues to the *unconscious* mind, because dreaming is thought to express levels of unconscious wish-fulfilment expressive of the individual's deepest conflicts and desires.

Dreamwork A term used by Freud to mean the complex process by which unconscious wishes and fantasies are disguised in dreams, appearing in symbolic form.

Drive An energized state in which the organism is motivated by the need to satisfy some (usually physiological) lack or want.

Drive-reduction theory The theory that motivation occurs, and behaviour is energized, mainly or entirely as a result of the need to alleviate or reduce drives. It is a rather negative theory in that it assumes that all drives produce tension or arousal and that the organism is always motivated to minimize drive states. The failure to encompass enjoyment and activities which deliberately increase arousal (like exploration and sky diving) was one reason for the decline of the theory.

Drug A chemical substance, usually non-nutritive, which exerts an effect on the body.

DSM IIIR An acronym for the Diagnostic and Statistical Manual (third version, revised) produced by the American Psychiatric Association. The Manual, from its first version, has been an attempt to standardize diagnosis. It is a useful way of finding out what is currently regarded as good practice in diagnosis. However the empirical evidence for its *reliability* and *validity* have been disappointing, although perhaps not less so than for any other psychiatric technique.

Dual-memory theory A model of memory first proposed by William James, in 1890, and later developed by (among others) Miller and Atkinson and Shiffrin. Dual memory theory postulates two independent memory systems, a limited-capacity, immediate or *short-term memory* (STM), and a large-capacity, *long-term memory*

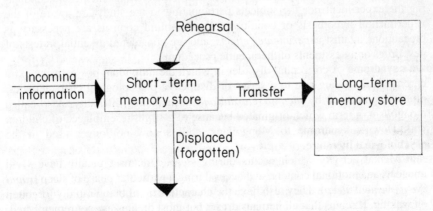

Dual memory theory

(LTM). The Atkinson and Shiffrin model proposes that STM forms a first stage to LTM storage, and that material is transferred from STM to LTM by means of rehearsal. See also *levels of processing*. [f.]

Dynamometer A machine for measuring muscular strength, particularly of hand grip.

Dysfunctional Working or operating in such a manner as to be a positive handicap to the individual or originating body. In general, the prefix 'dys' is used to mean faulty or presenting a problem with.

Dyslexia A general term for disorders involving a failure to learn to read, or specific difficulties in the interpretations of words or letters, despite adequate general intelligence. Dyslexic problems may take many forms, and there are a number of theories as to origins and therapies.

E

Eardrum The part of the ear which forms a barrier between the outer ear and the middle ear. The ear drum, or *tympanic membrane*, is a taut membrane which vibrates in response to sound. These vibrations are transmitted to the middle ear, where they are amplified, and then on to the cochlea for transduction into electrical impulses.

Eating disorders A general term for disturbed behaviour involving food. Includes *anorexia*, and *bulimia*.

Ecological validity *Validity* is concerned with the question of whether a given psychological technique really assesses that which it purports to measure. Ecological validity is, as its name suggests, concerned with whether a given technique truly corresponds to its equivalent in an everyday, 'natural', setting. The issue centres around whether artificially controlled, laboratory simulations of human situations can really be considered to be equivalent to the behaviour which human beings emit during the course of their everyday lives, given what we know about *demand characteristics* and *self-fulfilling* prophecies. So, for example, it is questionable whether *minimal group studies* of *social identification* are really examples of the same psychological processes as are produced by belonging to a given ethnic or occupational group, since they form highly confined and restricted laboratory studies which deliberately exclude all the complexities of social life and retain only a single, simple categorization as the distinctive feature determining social interaction.

Ecology The study of environments, with an underlying assumption that environmental characteristics are responsible for the ways organisms function.

ECT See *electroconvulsive therapy*.

Educable mentally retarded An American term corresponding to the obsolete British category of educationally subnormal, and implying a delayed mental development such that the child cannot cope with normal schooling, but is still educable if special help is provided. Specifically the category is applied to children with an IQ between 50 and 69. Below 50 the American term is trainable mentally retarded.

Educational psychology One of the major professions of psychologists. In Britain practitioners are employed within the educational system to deal with psychological issues concerning children in school, and to assess and monitor the progress of children with special needs. They are usually based in Schools Psychological Services or Child Guidance Clinics. In some areas the work is largely taken up with assessing children who are having difficulties in school and making recommendations about which kind of educational setting they need. Other areas have been able to develop much more varied work ranging from therapy with individual children and their families, through curriculum development and teacher training, to consulting with the school on more effective management structures. Training courses usually last for two years and award a master's degree, but require the applicant to have a good psychology degree, training as a teacher, and two years of teaching experience before starting the course.

Educationally subnormal A classification for children who are unable to cope with normal schooling. A Government Act (1981) ruled out the use of all of the terms which had been set up to label children with mental or physical handicaps, so these terms are now obsolete. When a child has 'special needs' a report is prepared by an *educational psychologist* which defines the strengths of the child and the areas in which special help will be required. It is hoped that this will prevent children being tied to *labels* which inevitably become derogatory and difficult to remove.

EEG The electro-encephalogram which is a recording of changes in the overall electrical activity of the brain. EEGs are taken by means of attaching several electrodes to different parts of the scalp, and using these to detect neural activity in the different regions of the brain. A *polygraph* converts these fluctuations into a written record and/or sends them for computer analysis, so that particular patterns of activity or responses to specific stimuli can be indentified. Specific frequencies (*alpha, beta* and *delta*) are reliably associated with different mental states, and patterns of EEG response can be used to identify a disposition to fits (seizures) and other forms of brain dysfunction. [f.]

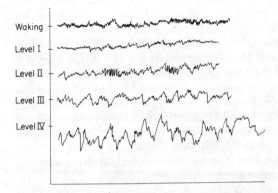

EEG recordings during different levels of sleep

Effect When used as a noun, an effect is a result or consequence ('the effect of his action'). When used as a verb, it means to bring about a consequence, e.g. I may effect a change in your understanding. This is an entirely different meaning from *affect*, and students need to be careful not to confuse the two, as the entire meaning of a sentence may be changed by such a mistake. (See *affect* for more detail).

Efferent neurone See *motor neurone*.

Effort after meaning A term used by Bartlett to describe the ways in which individuals attempt to organize their memories, and to make sense of them; if necessary, altering the content of the specific information in order to do so.

Ego In Freudian theory, the part of the personality structure that deals with external reality and controls the energies of the *id*. Literally the word means 'I' and its more general use is to refer to the sense of *identity* or *self*.

Egocentric speech Speech which is simply involved in monitoring and directing the child's internal thought processes and has no communicative function. According to Piaget, this is a significant part of the child's acquisition of speech; it forms a valuable tool of thought, which the child practises as it performs mental *operations* on the external world.

Egocentricity or egocentrism A central concept in Piagetian theory, egocentricity refers to the idea that children take their own perspective as central, and tend to assume that other people have the same understandings, motives and needs as the child. It is not a moralistic concept, and has nothing in common with selfishness or *egotism*; but instead is concerned with the child's perception of association and causality. The process by which the child gradually comes to differentiate itself from the external world, through the development of the *body-schema*; to recognize that objects have permanent existence; and to be able to *decentre* and see things from another's viewpoint are, for Piaget, significant milestones in the reduction of egocentricity. When used of adults the term does have implications of selfishness, though perhaps it should just imply a delayed cognitive development.

Ego-defence mechanisms See *defence mechanisms*.

Ego-ideal The image an individual holds of the person they would like to be. Also called the 'ideal self'.

Egoism A tendency to give an excessively high priority to one's own needs and wishes and a correspondingly low priority to those of other people. See *egotism* for a comparison.

Egotism A consistent tendency to overvalue oneself and therefore to undervalue other people. The difference from egoism is that egotists tend not to be interested in other people, whereas egoists need a good understanding of others in order to exploit them. Egotistical attitudes tend to be clearly displayed whereas egoism may need to be concealed to be effective.

Eidetic imagery Commonly referred to as *photographic memory*, eidetic imagery is memory which has been encoded by means of a particularly detailed visual image, such that the individual is sometimes able to identify details from the image which are unlikely to have been noticed on first exposure. Eidetic imagery is relatively common in children, occurring in about 10% of cases; but tends to disappear round about late adolescence, and is rare in adults.

Einstellung A term coined by the *Gestalt* school of psychology to refer to the kinds of mental *sets* which can influence problem-solving by inducing a rigidity of thought which precludes the perception of alternative strategies or solutions.

Elaborated code A term used by Bernstein to refer to the form of language commonly used by middle-class families, characterized by an extensive use of nouns, explanations, and synonyms. Bernstein's use of the term code is contentious, as are many other parts of his theory. This is due mainly to the theory having been associated with the *verbal deprivation* hypothesis of class differences in language use, which argue that restricted language use implies restricted cognitive possibilities. See *code of language, restricted code.*

Elaboration In cognition, the addition of information to a representation or *schema* that already exists in the cognitive system. This may involve incorporating new information (which Piaget called *accommodation*) or relating the representation to other stored information.

Electra complex A term introduced by Jung as a female counterpart of the *oedipus complex.* Most theorists, including Freud, have rejected use of the term.

Electrical impulse A short burst of electricity. In most psychological contexts this refers to the electrical impulse produced by the *neurone* when it is stimulated. See also *all-or-none principle, synaptic transmission.*

Electrocardiogram (ECG) A recording of the electrical discharges which appear on the surface of the body as a result of the activity of the heart. In psychology the ECG is used mainly as a way of recording the rate at which the heart is beating, as changes in *heart rate* may indicate the presence and strength of a number of processes such as *stress* and *orienting.*

Electroconvulsive therapy (ECT) A treatment for *endogenous depression* which involves passing an electrical current through the brain, thus simulating a severe epileptic fit. This controversial treatment induces temporary amnesia and often results in an alleviation of some forms of depression. However, there is considerable concern both as to how appropriately it is actually used in the psychiatric context, and with regard to the possibility of long-term damage to memory and concentration.

Electro-encephalogram See *EEG.*

Electroshock therapy American term for *electroconvulsive therapy.*

Electrode A device which will pick up or transmit electrical activity wherever it is placed. Normally it takes the form of a small metal disc, coated with a jelly to improve electrical contact and fitted to a larger adhesive disc so that it can be securely attached to the skin.

Emblems A category of non-verbal signals identified by Ekman & Friesen, which involves those non-verbal cues which have a direct and culturally understood meaning, which stand for something. Gestures with specific meanings, or uniforms denoting specific role functions might be examples of emblems.

Embryo An organism in the earliest stages of development following conception. In lower species the animal is called an embryo until hatching or birth. In humans the period of the embryo extends to two months from conception by which time different organs are becoming visible, and the term *foetus* is then used until birth.

Emergency reaction See *alarm response.*

Emergent properties Properties or characteristics which appear in groups or combinations of elements which could not have been predicted from the characteristics of the individual elements which make up that group. One of the chief arguments against *reductionism* as a form of argument is in its assumption that complex behaviour, whether it be social or individual, can be explained simply by reference to its component parts. This does not take into account the emergent properties which become apparent when elements are combined into a higher-order whole. So, for instance, it would not have been possible to predict that group decision-making can result in highly polarised decisions simply from looking at the decision-making patterns of individual group members (see *group polarisa-*

tion); or to have identified the phenomenon of *groupthink* from research into individual cognitive processes.

Emotion The experience of subjective feelings which have positive or negative value for the individual. Beyond this statement the definition must depend on the particular theory of emotion being held. Most current theories regard emotions as a combination of physiological response with a cognitive evaluation of the situation. The idea that emotions are the source of actions has become less popular and in fact the term has only a remote link with any idea of motion, having come into English from the French word emouvoir, 'to excite'. Some definitions would reserve the term emotion for fairly intense and fairly brief experiences. It is certainly useful to distinguish emotions from states, (like hunger, sexual desire, and frustration), which may give rise to emotions, and from behaviours such as aggression, which may indicate the presence of an emotion but which are not themselves emotions.

Emotional disorder A set of disorders in which children show high levels of *shyness, anxiety,* and *dependency*. The term is also used more broadly to refer to a wide range of psychological disturbances which involve inappropriate emotional experiences, such as *mania* and *depression*. For this sense an alternative term is *affective disorder*. See *conduct disorder*.

Empathy A feeling of emotional understanding and unity with another, such that an emotion felt by one person is experienced to some degree by another who is empathic to them. It is sometimes used in indicating how much capacity an individual has to be empathic towards others. It is thought to be important for psychotherapists to be empathic. See *WEG*.

Empirical Such as can be measured. Empirical observations are those which can provide a level of objective data which can be assessed in one form or another. Using the term 'measure' loosely, almost all psychological forms of investigation may be considered to be empirical.

Empiricism A philosophical school of thought highly influential in psychology, which argued that only that which can be directly observed or measured can be meaningfully studied.

Enactive representation According to Bruner, the first *mode of representation* developed by the young child. Enactive representation involves the storing of information in the form of *kinaesthetic* sensations, such as the way that most adults would recall the sensation of a fairground waltzer or helter-skelter. In the world of the infant, such 'muscle memories' would be adequate to cope with most of the information encountered by the child; as the child develops and its world widens, further forms of representation are added to its repertoire, such as *iconic representation* and *symbolic representation*.

Encoding The processing of information in such a way that it can be represented internally, for memory storage. The term is also used when data are transferred into a standard form such as a computer file.

Encounter group A therapeutic technique devised by Carl Rogers, in which clients are placed in a situation which facilitates openness and honesty about their self-concept and feelings concerning the others in the group. Once initial barriers were down, Rogers considered that such a group would provide the emotional support and *unconditional positive regard* needed for each member to deal with their problems and to explore their options for personal growth. See also *client-centred therapy*.

Endocrine system This is the general term given to a system of glands distributed throughout the body, which release *hormones* into the bloodstream. The endocrine system is generally involved in the maintenance of specific conditions of the body, such as pregnancy or aroused states; rather than in particular acts or behaviours. The main gland of the endocrine system would appear to be the *pituitary gland*, which is located in the brain and directly connected to the *hypothalamus*. The pituitary sends messages to many of the other glands of the system, and is closely linked to the homeostatic mechanisms of the body. Some of the other glands of the endocrine system are the *pineal gland*, the *thyroid* and *thymus glands*, involved in growth regulation and immune mechanisms; the *testes*, and the *adrenal glands*.

Endogenous depression *Depression* which has come about without any apparent cause, and which persists over an extended period of time. The term implies that the depression originates within the individual, rather than being a response to external circumstances. See also *reactive depression*.

Endorphins These are a group of neurotransmitters mainly found in the limbic system, which were originally termed 'endogenous morphine'. This later became contracted to endorphin, and the group includes the similar *enkephalins*. They are substances with chemical structures closely resembling morphines, produced in the brain in response to demanding exercise, pain, anxiety or fear. It is considered that the feelings of euphoria which often accompany strenuous exercise are produced by the actions of endorphins; and that the similar experience produced by the opiates *heroin* and *morphine* occur as a result of their being picked up in *receptor sites* specific to endorphins.

Engineering psychology The application of psychology to *man-machine interaction*. It includes the selection and training of people to operate machines, and advice on the design of machines so that they can be efficiently used by human operators.

Enkephalins A specific set of neurotransmitters belonging to the general group of *endorphins* which are produced in response to pain or demanding exercise.

Environment The total external context in which an individual operates. The concept of environment is usually used to include physical surroundings and their characteristics and social contexts and interactions, but it may be used more specifically to include all the different facets of the physical but to exclude the social. See *ecology*.

Environmental determinism The view that behaviour, personality, or psychological characteristics originate as a direct consequence of individual learning and environment influences, and are not significantly influenced by *innate* factors.

Environmental psychology The study of the ways that the environment influences and channels individual behaviour. Environmental psychology includes the study of such factors as *territoriality* and *personal space, ergonomic* design, and the physical attributes of surroundings.

Environmentalism The doctrine that all significant determinants of behaviour are to be found in the environment. Strict behaviourism is one version of environmentalism. See *heredity–environment controversy*.

EPI The Eysenck Personality Inventory; a questionnaire designed to assess people on the two character *traits* of *extraversion* and *neuroticism*. These were proposed as the two main traits underlying *individual differences* in *personality*; each representing several second-order traits.

Epidemiology A research technique in which the distribution of the events or other features under study is plotted, in order to identify any patterns or regularities. Distributions may be plotted geographically, e.g. studying whether identified cases of incest are more prevalent in some areas of a city than others. Alternatively, other forms of distribution may be applied, such as in epidemiological studies of the incidence of AIDS in subgroups within the *population*.

Epilepsy A physiological disorder which involves a rapid and unpredicted electrical discharge, starting in the left temporal lobe of the brain and spreading across the cortex. Such a discharge produces temporary changes in mental state, uncontrollable muscular spasms (a fit, or seizure), and leaves the individual with temporary *amnesia* and often feeling very tired. There are many degrees of epilepsy, and although a few unfortunate individuals experience *grand-mal* attacks, which involve a temporary loss of muscular co-ordination and prolonged tembling, many epileptics show little outward sign of the attack. Epilepsy has historically been regarded with a high degree of superstition, but, apart from practical difficulties of lifestyle encountered by epileptics, does not affect the person between attacks. It is difficult to suppress, and anti-epilepsy drugs tend to have strong and unwanted side effects. In the most severe cases, *split-brain surgery* has been attempted with some degree of success.

Epinephrine The American name for *adrenaline*.

Episodic memory Memory for specific events, episodes, or phenomena. See also *semantic memory*.

Epistemology The study of knowledge, and the ways in which what counts as knowledge may vary from one discipline or field to another. For example: a similarly worded question on, say, the family might occur in both a sociology and a psychology examination, but different epistemological demands would be applied in the evaluation of the answer. In the sociology exam, the student would be required to consider the relationship of the family to society; while in the psychology exam an examination of interpersonal process and roles would be more appropriate. What counts as knowledge in each subject is different, and it is the consideration of such differences that is the subject matter of epistemology.

Equal-interval scale A system of measurement in which the difference in value between consecutive units is consistent throughout. For example, in an equal-interval numbering scale, the difference between 30 and 31 is of exactly the same magnitude as the difference between 36,005 and 36,006. Equal-interval scales with a fixed zero are known as ratio scales. Children are introduced to ratio scales from a very early age, as they first learn to count; and adults continue to use them, e.g. in dealing with money. See *levels of measurement*.

Equilibration In Piagetian theory, the process by which *schemata* are developed to take account of new information. If new information which is encountered fails to fit into an existing schema, the individual is thrown into a state of cognitive discomfort, known as disequilibrium. Through the two processes of *assimilation* and *accommodation*, the schemata are adapted or adjusted such that the new information can be handled, and the cognitive balance is restored. This is the process of equilibration.

Equilibratory senses The *kinaesthetic* senses based on receptors in the *semi-circular canals* of the inner ear; such as the senses of balance and *proprioception*.

Equipotentiality The principle outlined by Lashley after his investigations of the cerebral cortex, that those areas termed 'association cortex', concerned with learning and memory, were equal in their potential to carry out these functions. In other words, that such functions were not localized, but organized across the whole of the association cortex. See also the *Law of mass action*.

Equity theory The idea, in accordance with social exchange theory, that people choose relationships in which they will benefit to about the same extent as they contribute. There is some evidence that if people feel either disadvantaged or over-advantaged in a relationship, they will be dissatisfied.

Ergonomics The study of the relationship between energy expenditure and work. As such, ergonomics includes the study of design and physiological limitations, and of other factors influencing efficiency in both mechanical and man-machine systems.

ESB The usual abbreviation for a form of direct electrical stimulation of the brain which appears to function as a powerful reinforcer of behaviour, and to give highly pleasureable sensations. Experiments conducted in the 1960s seemed to imply that there was a direct 'pleasure centre' in a particular region of the *hypothalamus*. For instance, stimulation of this area in rats, given as a reward for lever-pressing, produced an extremely high response rate; and in terminally ill cancer patients produced reports of feeling 'wonderful' or 'happy' (Campbell). It was thought that this might be the root of all motivational states. However, the 'pleasure centre' concept presents some difficulties; for instance, unlike other forms of learning, it extinguishes very quickly, so the status of ESB is now rather unclear.

ESP See *extrasensory perception*.

EST Abbreviation of *electroshock therapy*.

Esteem needs One level of the *hierarchy of human needs* proposed by Maslow. Esteem needs include the need for achievement and social recognition; and are considered to achieve importance once *physiological, safety*, and *social needs* have been met. See also *self-esteem*.

Esthetic See *aesthetic*.

Ethical To do with rights and wrongs. Owing to the scope of psychological interests and the potential for psychological damage, ethical issues have become of great importance in modern psychology. They include such aspects of psychological practice as the use of deception in experimental work; the investigation of characteristics which are potentially threatening to the self-concept (c.f. Milgram's work on obedience); the use of animals in research; and questions of confidentiality in professional practice. Professional psychological associations usually have special committees which evaluate and provide guidance on ethical issues.

Ethnocentricity A condition in which the perceptual framework and social assumptions of an individual are entirely bounded by, and defined in terms of, the experience of their own social, ethnic or national group. Ethnocentricity is therefore a form of cognitive (or rather, socio-cognitive) *set*, which leads to assumptions about one's own group's practices, beliefs or assumptions as setting the standard of 'rightness' or objectivity, and thus leading to undervaluing, or even failing to recognize, alternatives. Probably deriving from mechanisms of *social comparison*, ethnocentricity appears to be a fundamental and extremely common aspect of human thinking. It is clearly recognizable when we are looking at the arguments of those belonging to different social groups, although difficult to recognize when we are looking at our own. Arguably the most powerful benefit of education, travel, contact with others of different backgrounds, etc. is that it can sometimes have the effect of reducing the extent of the individual's ethnocentricity. Regrettably, however, this is not an inevitable consequence of any of these experiences.

Ethogenics An approach to social enquiry outlined by Rom Harré (Harré, 1979), in an attempt to identify some of the more meaningful aspects of social interaction. Properly speaking, ethogenics is a philosophy rather than a methodology, but there are two outstanding methodological implications of the ethogenic approach. The first of these is that it is the episode, rather than the act or action, which should constitute the basic unit of social enquiry, since social life is experienced in real life as a succession of meaningful episodes. Harré suggests that a dramaturgical metaphor may be helpful in episode analysis – if an episode is thought of as similar to an act in a play, then a number of features of the situation become significant in interpreting it: characters, setting, scripts, non-verbal communication, prior episodes, plot and so on. Analysis of these different aspects of the episode would therefore link diverse areas of psychological knowledge to provide an insight into what is going on.

The second methodological implication of the ethogenic approach is that the accounts which people give of their experiences should be taken to have equal validity to an external 'objective' analysis, since the way that we perceive and experience social life is just as important in determining social interaction. *Account analysis*, in Harré's model, has two stages: the first being the process of collecting the accounts themselves, and the second consisting of a critical reflection of the meanings contained in those accounts. See also *new paradigm research; qualitative analysis, emergent properties.*

Ethology The study of behaviour in the natural environment. Ethological studies of animal behaviour have been conducted throughout the 20th century, and were systematized by the work of Konrad Lorenz and Niko Tinbergen. More recently, the ethological approach has been applied to the study of human behaviour, most notably in the fields of *mother–infant interaction* and *non-verbal communication*.

Eugenics A set of political beliefs based on the idea that intelligence and personality are fixed inherited characteristics determining role and position in society. Eugenicists believe that breeding should be restricted among those of the 'lower' classes of society, and that those of subnormal intellect or undesirable personality should be sterilized to prevent the spread of such genetic characteristics. Eugenic ideas were widespread in Western Europe and America before the Second World War, mainly as a result of the work of Francis Galton, and formed the basis of the Nazi policy of 'exterminating' those considered to be of inferior racial characteristics. Eugenic laws were also enacted (and in some cases are still current) in several states of the USA, and there are many cases on record of individuals classified as mentally subnormal having been involuntarily sterilized as a result of these laws.

Euphoria Extreme happiness; a feeling of being elated or 'high'.

European Social Psychology A school of thought in social psychology which derived from theories developed by European psychologists, particularly through the 1970s and 1980s. One of the central theories in this approach is that of *social identity theory*, particularly associated with the work of Henri Tajfel (e.g. Tajfel, 1982). Social identity theory is concerned with how people internalize social group membership and interact as representatives of their social group rather than as individuals. Another core theory in European Social Psychology is *social representation theory*, developed by the French psychologist Serge Moscovici (e.g. Moscovici, 1984). This is concerned with the shared beliefs which emerge in society and which serve to legitimize and rationalize social action. A third area of interest is research into social and collective attributions, for example in the work of Miles Hewstone (e.g. Hewstone, 1989). European social psychology can therefore be perceived as a body of theory spanning several different levels of explanation in social life. It emerged particularly with the formation of the European Journal of Social Psychology in the early 1970s, and has been proposed as a marked contrast to the largely problem-centred, and some say reductionist, approach represented by much of American social psychology. See also *social attribution*.

Evoked potential A measure of brain activity obtained by taking an *EEG* reading at the same time as exposing the individual to some form of stimulation – usually visual. The resulting changes in the EEG record are known as the evoked potential. In practice the stimulus is usually applied repeatedly and the responses averaged so that the signal can be distinguished from the background noise.

Evolution A gradual process of genetic change in which the genetic characteristics of a whole species are altered over many generations, effecting a physical change which serves to adapt the individuals of that species more fully to their environment. Individuals in a species do not change, but owing to the genetic re-shuffling which occurs as a result of *sexual reproduction* (or to mutation) each individual varies genetically from its parents. If the variation is one which confers an advantage, in terms of the adaptation of the animal to its environment, then that individual is likely to become stronger and healthier, or in some other way more likely to breed and to pass on its favourable genetic characteristic to its offspring. Gradually, over time, weaker members of the species are less efficient at surviving, and so the 'new' genetic characteristic becomes more widespread in the population. This process is known as natural selection. Over millions of years, this results in the development of whole species specialized to their environment. Although evolutionary arguments are frequently voiced in terms of modern humankind, these are unlikely to have much substance, owing to: (1) the relatively few generations involved in 'modern' lifestyles; and (2) the tendency of humankind to modify its environment to suit itself, thus obviating the need to alter the species to suit the environment. See *coevolution, sociobiology*.

Excitation The process by which a *neurone* is rendered likely to fire. Excitation of neural impulses occurs either through direct stimulation of sensory neurones from sense receptors receiving information from the environment, or through the stimulation of a number of *excitatory synapses* making connections with that particular neurone.

Excitatory synapse A *synapse* which, when stimulated, renders the neurone receiving the neurotransmitter more liable to generate an *electrical impulse*. Although stimulation from a number of excitatory synapses is usually required to set off the nerve impulse, reception of the appropriate *neurotransmitter* serves to lower the *threshold of response* of the neurone, thus contributing to the eventual production of the impulse.

Existentialism A philosophical approach which argues that individuals can only be understood in terms of their existence in the world and the choices with which they are faced. Existentialists stress self-determinism rather than environmental or developmental determinism, and emphasize the responsibility which each individual has for his or her actions within society, on the grounds that the individual is always free to act differently, to say 'no', and to accept the consequences. Existentialism was extensively propounded by Jean-Paul Sartre, and has been taken up by many psychological theorists. Most notable of these was probably R.D.

Laing, who in *The Divided Self* proposed an existentialist theory of *schizophrenia* which directly challenged orthodox psychiatric approaches and stimulated investigation of several alternative forms of therapy in cases of psychological disturbance, such as *family therapy*.

Exogenous Outside the person. Compare *endogenous*.

Exogenous depression A depression that is believed to be caused by external events. Usually called a *reactive depression*.

Experiment A form of empirical investigation or study in which variables are manipulated in order to discover cause and effect. An experiment will involve at least one *independent variable*, which will be set up in such a way as to produce changes in a *dependent variable*.

Experimental design The process by which an experimental study is organized so as to allow for investigation of the possible effects of the *independent variable* upon the *dependent variable*, with as little contamination as possible by *confounding variables*. See *counterbalancing, experimenter effects, matching*.

Experimental group A sub-group of the subjects in an experiment who all receive the same version of the experimental condition. When there is only one experimental condition the experimental group is compared with the *control group*. In more complex designs there will be several experimental groups, each receiving a different condition. In these cases the subjects who receive any kind of experimental condition may be called the 'experimental subjects'.

Experimental method The use of controlled experimental situations to test *hypotheses*. The term is rather vague largely because there is no agreed definition of what constitutes an *experiment*.

Experimental neurosis Laboratory studies can induce apparently neurotic behaviour in animals by training them to perform a task and then gradually making it impossible. First studied by Pavlov and presented as a basis for controlled study of *neuroses* in humans. Subsequently doubts were raised about whether the mental states of the animals were really similar to those of neurotic humans, and the research was abandoned. A similar process occurred more recently with the study of *learned helplessness*.

Experimental philosophy That branch of philosophy which, during the 18th and 19th centuries, became increasingly concerned with the study of the human mind, and which drew on *empirical* observations for its conclusions. Experimental philosophy became transmuted into psychology towards the end of the 19th century; the 'founding fathers' of psychology, Wilhelm Wundt, Herman Ebbinghaus and William James were simultaneously the last of the experimental philosophers.

Experimental psychology Those branches of psychology which are firmly based in laboratory experimentation. The term was used to cover such areas as learning, memory and perception. It has now been largely replaced by the wider area of *cognitive psychology*.

Experimenter effects Experimental problems producing a biased result brought about by the influence of an experimenter, for example, through subjects responding to the person who conducts the experiment. Experimenter effects may occur indirectly, because of the personal characteristics of the experimenter (e.g. their age, sex or other such feature), or directly, as a result of the beliefs or unconscious bias being transmitted to the subjects, and producing a self-fulfilling prophecy. The latter is usually controlled by using the *double-blind technique*. See *demand characteristics*.

Exploration Activity undertaken to gain information. Vigorous exploratory behaviour is characteristic of young in many species and is often studied in conjunction with play. Daniel Berlyne proposed a major distinction between diversive exploration, in which the environment is investigated to identify sources of possible interest, and specific exploration, in which attention is focused on a specific object or phenomenon. Specific exploration is usually more systematic in investigating the properties of the object. Some research has investigated which properties of objects are most likely to elicit exploration but a major reason for being interested in exploration is that it appears to originate largely within the child or animal. It is therefore a potential source of self-motivated learning.

External attributions See *internal attributions*.

Extinction A term used in both *classical* and *operant conditioning* to refer to the dying-out of a response as a result of lack of reinforcement. In *behaviour therapy*, learned associations, such as phobias, are treated by procedures designed to effect extinction of the stimulus–response connection, by organizing circumstances in such a way that it will not be reinforced. [f.]

Extrasensory perception (ESP) Perception which does not depend on the usual

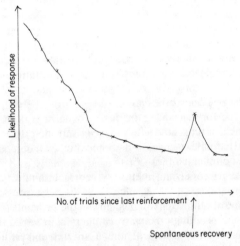

The extinction of a learned response

sensory processes. ESP is one of the class of *paranormal phenomena* and, in common with other forms, there is a lack of clear evidence that it actually happens. There is as well the rather odd scientific status that, if it could be explained in terms of known laws, it would cease to qualify for the title. See *parapsychology, telepathy*.

Extraversion A term originally coined by C.G. Jung, to describe individuals who are outward-directed and sociable in their behaviour. Extraversion as a personality trait was adopted as one of the main personality dimensions by Eysenck, in his two-factor model of *personality*, in which it included such second-order factors as risk-taking, *impulsiveness*, and *sociability*. See *EPI*.

Eye contact Mutual gaze, or the amount of time which two people spend looking at each other simultaneously. Eye contact is sometimes taken as an indicator of intimacy; eye contact with unknown individuals tends to be avoided. It is a powerful signal in all primates, including human beings: prolonged eye contact with neutral or hostile facial expression is taken as a threat gesture, and tends to be responded to by either aggressive or avoidance behaviour.

Eyebrow flash A recognition signal which consists of rapidly raising the eyebrows as a greeting to an individual who is recognized. The eyebrow flash seems to be common to all human cultures, and to some other species. It therefore is considered to be *innate*.

F

Face validity See *validity, face*.

Facial affect programme A strategy of inducing behavioural change through making the individual aware of the sensations arising from facial expressions which are different from those that he/she uses habitually. It is thought that encouraging the continued use of positive facial expressions, as opposed to those normally used, will provide positive feedback both through social interaction and through muscular interpretation. See *facial feedback hypothesis*.

Facial electromyography A technique for measuring the degree of tension in facial muscles by recording the electrical discharges of the muscles. By mapping the muscle tensions occuring in different expressions a systematic and objective measure of facial expression can be obtained.

Facial expression Characteristic patterns of arrangements of the muscles in the face, which provide important non-verbal cues in social interaction. Facial expression may be used either to express understanding, attitudes, emotions, or as specific cultural signals with clearly defined meanings. Some researchers have found that basic emotional expressions seem to be common to all human cultures, and are also found in blind babies, which would seem to imply that they are innate. However, other facial expressions show cultural variability, and seem to be acquired through social interaction.

Facial feedback hypothesis The idea that our experience of emotion arises at least in part from our interpretations of the arrangement of our facial muscles. So mood changes may be effected by the altering of the facial expression, which will provide feedback leading to a change in the emotion that that person experiences. The effect is used in studies of mood when subjects are asked to make, say, a depressed face as part of a procedure for changing their mood.

Factor analysis A statistical technique much loved by psychometricians, which involves the analysis of large and complicated sets of data in such a way as to draw out the underlying pattern of correlations. Groups of measures which all inter-correlate are identified as a 'factor' and the researcher can then examine the measures to see what they all have in common, and then speculate about the cause of or reason for the grouping. The technique requires a large amount of calculation and is now invariably carried out by computer.

Failure to thrive (FTT) A condition of poor growth in infants, usually defined as being below the third centile (i.e. in the bottom 3% for that age, sex, and population). In some cases there is a physiological problem which accounts for the poor growth but in the majority of cases there is no organic cause and the condition is called 'non-organic failure to thrive'. FTT was once believed to be a direct result of emotional deprivation and in its extreme form was called 'deprivational dwarfism'. It is now widely recognized that the basic problem is that the child does not receive enough food to sustain appropriate growth, though this in turn is likely to result from emotional or other difficulties of the parent, the child, or both. See *child abuse*.

Family therapy An approach to psychological treatment in which the whole family is the focus, rather than an individual patient. Earlier approaches were derived from *psychoanalysis* and treated the family as if it had psychological processes similar to those of individuals. Recently, methods have been developed from *systems theory*, which recognize that, while the behaviour of a component may seem strange when it is seen in isolation, it will make much more sense in the context of the complete system. Applied to individuals, and recognizing that families are one of the most significant systems within which most people function, this approach has led to a new way of looking at psychological disturbance. It assumes that in many cases the 'symptoms' shown by an individual are a meaningful response to their circumstances. More specifically, disturbed behaviour is likely to be an attempt to regulate relationships, or solve problems, within the family. The literature contains many examples of spectacular success using 'systemic family therapy' but there has been little systematic evaluation of the techniques.

Fantasy The conscious mental construction of images of events or objects. Generally a pleasurable activity which may indicate psychological health, and may be useful in creatively exploring possible courses of action. The content of fantasy, like dreams, may reflect major unresolved conflicts, and an excessive investment in fantasy indicates psychological problems. See *daydreams*.

Fatigue effect An experimental effect brought about by the subject's being tired, bored, or otherwise affected by the duration of the experimental procedure. It can contaminate experimental results because it may appear that subjects are less good at later tasks when in fact they are just getting tired. See *counterbalancing, order effects*.

Fechner's law A principle in *psychophysics*, which states that the sensation experienced by an individual increases as a logarithmic function of the stimulus intensity. In other words, that the physical increase in stimulation required for a perceived increase in intensity is not constant, but systematically greater for higher intensities. For example switching a light on may be perceived as a substantial increase in brightness when the room was previously dark, but may be hardly noticeable during bright sunlight. See also *Weber's Law, just noticeable difference, relative threshold, absolute threshold*.

Feedback Information which informs the individual about the effect or outcome of a course of behaviour which has been enacted by that person, thus allowing a sequence of actions or behaviour to be modified if necessary or desirable. See *biofeedback, negative feedback*.

Field dependence/independence An aspect of *cognitive style* concerned with whether a person is dominated by context when making judgements (field dependence) or whether they can ignore distracting contextual information (field independence). It may be tested by the accuracy with which a subject can judge the orientation of a line when it is surrounded by a frame at a different angle, or when the subject is in a chair which can be tilted away from the vertical. Large *individual differences* have been found, which seem to relate to other areas of cognitive functioning.

Fight or flight response See *alarm reaction*.

Figure–ground organization The tendency, which is built into our visual perception, to organize incoming information (which arrives in the form of light waves of varying intensities and wavelengths) into meaningful units, or figures, set against a background. Figure–ground organization was intensively studied by the *Gestalt* psychologists, who identified several principles of perceptual organization which served to make up figure–ground discrimination. These were collectively known as the *Laws of Prägnanz*, and included the principle of closure, and the principle of 'good gestalt'. [f.]

Figure–ground perception This is a general term used to refer to those aspects of perception which derive from *figure–ground organization*. So, for instance, it would include areas such as *pattern perception*, which is dependent upon the organization of visual information into figures against background.

Rubin's vase

Rubin's vase: an example of alternating figure and ground.

Filter models Theoretical models put forward to suggest plausible mechanisms by which cognitive processes may take place. The best known filter models were put forward to explain the process of *selective attention*, by psychologists such as Broadbent, Triesman, and Deutsch. Each of these represented a more or less complex attempt to explain the way in which incoming information is channelled such that only a selected part of it is received, rather than the overwhelming whole. [f.]

Fixation In psychoanalytic theory, the failure to progress from an earlier stage of development, (e.g. oral fixation) or an earlier relationship (e.g. mother fixation). The term is also used more broadly for any relationship which is seen as inappropriately intense and dependent.

Fixed action pattern Complex sequences of behaviour that are genetically pre-programmed so that all members of the species show the behaviour when it is

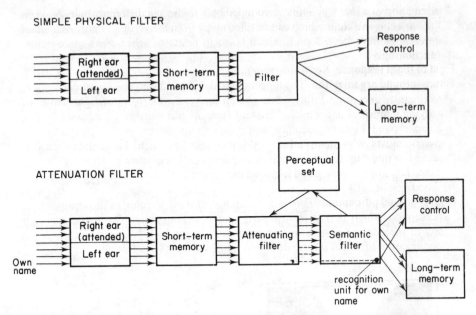

SIMPLE PHYSICAL FILTER

ATTENUATION FILTER

Filter models of selective attention

needed. Neither learning nor practice are needed for the behaviour to be performed perfectly. Fixed action patterns have been intensively studied by *ethologists*, and involve sequences of behaviour which have been inherited as a complete unit.

Fixed-interval reinforcement A *reinforcement schedule* in which reinforcements, or rewards, are given only after a set period of time since the last reinforcement became available. After a suitable *acquisition* period, this method of administering reinforcement tends to produce a high level of responding around the time of the reinforcement, and a low rate of responding at other times. It has a low *resistance to extinction.*

Fixed-ratio reinforcement A *reinforcement schedule* in which reinforcements, or rewards, are given only after a set number of responses has been made since the last reinforcement. Fixed-ratio reinforcement schedules produce a very rapid *response rate*, but have a low *resistance to extinction.*

Fixed-role therapy A method of treatment derived from *personal construct theory* in which the client agrees to adopt particular ways of behaving which are clearly different from (though not opposite to) his or her usual style. The method seems particularly effective in undermining a belief that only one kind of behaviour is possible.

Flooding See *implosion therapy.*

Focal therapy An approach to psychotherapy in which a specific focus (problem) is identified early in the therapy and efforts are concentrated on this focus for the remainder of treatment. The method was developed as part of the attempt to make *psychotherapy* shorter and more cost-effective.

Foetus An organism in the later stages of gestation, up to the time of birth. In humans, from the ninth week after conception. Before this time it is called an *embryo.*

Forensic psychology The application of psychology to legal matters. Includes work on reliability of witnesses, evidence given by children, the consequences for children of possible court actions, and the causes of criminal behaviour.

Forgetting Broadly speaking, theories of forgetting can be sorted into seven major approaches: decay theory (the idea that memory traces gradually decay over time, unless strengthened by being retrieved); *interference* theory; *amnesia* brought about through physical causes; *motivated forgetting*; lack of appropriate *cues* for retrieval; lack of the relevant context for retrieval; and inadequate processing during storage (see *levels of processing theory*).

Formal operational stage The last of Piaget's four stages of *cognitive development*. In the formal operational stage, the individual has become capable of abstract thought and can conceptualize possibilities which are outside of direct experience. Piaget considered this to be the highest form of cognitive activity, and one which is shown only in human beings, and from the age of about 12 years at the earliest. The preceding stages he viewed as steps towards this point, which, on the basis that *ontogeny* recapitulates *phylogeny*, illustrated the stages by which abstract logic must have evolved. See also *sensori-motor stage, pre-operational stage, concrete operational stage, genetic epistemology*.

Fovea That part of the retina which receives the clearest and most sharply focused image. The fovea is roughly central to the retina, and is composed entirely of *cone cells*, which, through their discrimination of colour, allow for the distinguishing of fine details, at the cost of some loss of sensitivity to faint signals. See also *rod cells, retina*.

Fraternal twins Twins which have developed from two separate ova, such that they bear the same resemblance and relationship to each other as normal *siblings*. Twins of this kind are also known as *dizygotic* twins. See also *monozygotic twins, identical twins*.

Free association A technique much utilized by Freud and subsequent *psychoanalysts*, as they considered that it provides important clues to the workings of the *unconscious* mind. Free association consists of the individual producing an uncensored, non-calculated account of what they are thinking and feeling during the session. Because the spontaneous expression avoids intervention and possible censorship by the *ego*, (the conscious mind), the nature of the responses made during a free association session indicate the concerns and preoccupations of a person's unconscious. The agreement to engage in free association is called the 'basic rule' of psychoanalysis and is regarded as essential for its success.

Frequency theory A theory concerning the ways in which information contained in sound waves is transmitted to the brain. The frequency theory approach states that the wavelength, or frequency, of the sound affects the rapidity of transmission of electrical impulses along the auditory nerve, with sounds at higher frequencies producing more rapid transmissions. When the frequency rises to a point which would require firing at a more rapid rate than the neurones concerned can manage, the *volley principle* comes into effect, with neurones taking it in turns to fire, producing bursts, or volleys.

Freudian slip A mistake which can be interpreted as revealing unconscious wishes, fears, etc. Freud argued that all apparently accidental happenings reveal something of the unconscious.

Frontal lobe The general term given to the front part of the brain, located above the *lateral fissure* and in front of the *central sulcus*. In the early part of the century, the frontal lobe was thought of as the seat of aggression, from the discovery made by Moniz, in 1930, that chimpanzees which had experienced lobotomy – the surgical removal of the frontal lobe – showed a decrease in aggressive behaviour. This led to considerable popularity for lobotomy as an operation to treat those with *psychotic disorders*. The discovery that similar results could be achieved by the severing of connections between the frontal lobe and the rest of the cortex just above the lateral fissure (leucotomy) led to an equal popularity for the latter operation. However, it transpired that many other functions were also impaired, including, generally, the capacity for autonomous functioning and decision-making. Although the frontal lobe has few localized functions, it seems to be involved in much generalized cortical activity. See also *parietal lobe, temporal lobe, occipital lobe*.

Frustration Both the act of preventing an organism from reaching a goal, and the emotion aroused in the organism by the experience.

Frustration–aggression hypothesis The proposal, particularly associated with Leon Berkowitz, that aggression is always caused by some kind of frustration. It also tends to be assumed that frustration always leads to aggression. This theoretical model has achieved widespread popularity, and is supported by *comparative* studies of overcrowding in animals as well as by studies of human behaviour.

Functional fixedness A form of *einstellung*, or *mental set*, in which the individual is unable to deviate from using objects in a manner consistent with their normal functioning. So, for instance, in a problem-solving exercise, functional fixedness may prevent someone from realizing that something like a jug, usually used to contain liquids, could also be turned upside down and used as a support. See *problem-solving*.

Functionalism The claim that psychological phenomena are best understood in terms of their functions rather than their structure – which would be the claim of structuralism. Concepts such as *adaptation* and *role*, and therapeutic methods such as systemic *family therapy* represent a functionalist approach.

Fundamental attribution error See *attribution error*.

Future shock One of several theories about the stress imposed by *transitions* and *life events*. The idea was introduced in a book with that title by Alvin Toffler to describe what he claimed were the traumatic effects of our present rapid progress into the future. Toffler proposed that people could be protected against the effects of change by maintaining some areas of stability in their lives.

G

G See *general intelligence factor*.

Galvanic skin response (GSR) Also known as galvanic skin resistance, this is a highly sensitive measure of *arousal*, registering even such slight increases in arousal as are produced by a disturbing thought or a slight pain. It refers to the electrical resistance of the skin, which changes as a result of increases in the rate of sweating. GSR detectors form an important component in *polygraphs*, which record a range of physiological indicators of psychological events, and may be used as lie-detectors.

Gambler's fallacy A belief that if a chance event occurs, then it is less likely to occur on the next trial. If red comes up several times running on a roulette wheel there is a (mistaken) tendency to believe that black is more likely on the next throw. This universal tendency has been of interest to cognitive theorists as it is a failure to follow probabilistic logic and so may shed light on how humans assess probability. It may best be seen to reflect the fact that genuine instances of 'random sampling without replacement' are uncommon in real life and not as a failure to judge probabilities accurately. The gambler's fallacy is therefore a normally effective strategy which becomes inappropriate in certain, rather artificial, circumstances.

Game The psychological uses of this term are similar to the ordinary meaning except that the idea of playfulness is usually absent. So a game is an activity within defined limits in which all of the participants operate according to agreed rules. Much of social interaction can be regarded as a game, with plenty of scope for problems when the rules and the limits of the game are not made explicit. Eric Berne was one of the first to explore this concept in his book *Games People Play*. 'Game theory' is a specific approach which expresses the rules of the game in mathematical terms so that the possible strategies can be precisely identified and their consequences predicted. See also *zero-sum game*.

GAS See *general adaptation syndrome*.

Gaussian distribution See *normal distribution*.

Gender identity The awareness a person has of themselves as a member of their sex. It emerges from the relationships between the belief the individual has about appropriate sex roles, and their perception of themself. For example, a small man who regarded size and muscularity as indicators of masculinity would modify his gender identity accordingly.

Gene A unit of heredity, which consists of a small segment of a *chromosome* made up of DNA (*deoxyribonucleic acid*). Each gene exerts its influence on the body by triggering off a protein synthesis – usually the production of an enzyme, but sometimes a protein which goes towards the production of a particular type of cell in the body. The word gene is also used, loosely and erroneously, in sociobiological theory to mean 'a unit of natural selection'.

General adaptation syndrome (GAS) A long-term response to stressful stimulation identified by Selye in 1949. It is characterized by extremely high levels of adrenaline in the bloodstream, but without the rapid heart and pulse rates normally associated with *adrenaline* release, and changes to internal organs. Selye's research indicated that the effects are always the same, regardless of the source of the *stress*. GAS has been shown to result in increased susceptibility to illness, possibly through a decline in the number of white blood cells and antibodies produced by the body.

General intelligence factor (g) The idea of one overall capacity of *intelligence* suggested by Galton and Spearman. Many psychologists consider this to be a contentious view, arguing that intelligence is a combination of many differing skills and attributes. Most intelligence tests are based on the assumption that a generalized intelligence factor, or 'g', can be calculated as a result of the administering of a set of specialized sub-tests, and it is as a consequence of this belief that the *Intelligence Quotient*, or IQ, has been so widely applied. See *triarchic intelligence*.

General problem-solver (GPS) A computer programme devised in the early 1970's, which emphasized the use of *heuristics* in tackling specific problems; and which formed the prototype for many subsequent attempts at *computer simulation*, within the general field of *artificial intelligence*.

Generalization The process by which a learned response will occur in more situations than those in which it was first learned: it will also be applied to similar situations.

Generalization gradient The relationship between the strength of a given response and the similarity of the triggering stimulus to the original stimulus. When eliciting a generalized response, a stimulus which is very similar to the original will produce a strong response, while one which is less similar will evoke a weaker response. [f.]

Genetic(s) In the singular, concerning the origin of something. Used particularly to refer to the development of abilities and characteristics of children (see *genetic*

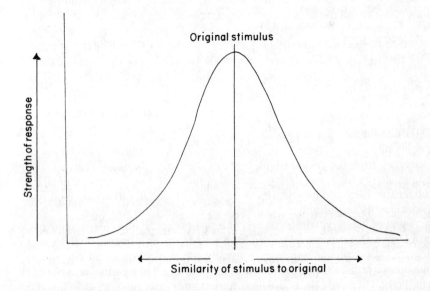

A generalization gradient

epistemology) but also applies to the development of characteristics in a species or the development of the species itself. See *ontogeny* and *phylogeny*. In the plural, genetics refers to the study of *genes* and their actions. See *behaviour genetics*. See also *Lamarckian genetics, Mendelian genetics*.

Genetic engineering The process of altering genetic characteristics through micro-scopic surgical or chemical intervention; usually taking the form of inserting a new section of *chromosome* into an existing one, such that when the chromosome is replicated, the new portion is also replicated and becomes part of the organism's overall *genotype*.

Genetic epistemology The title for a theory of the growth of knowledge and understanding. It is usually reserved for Jean Piaget's theory charting the development of the child's cognitive functioning through a series of stages. See also *Lamarckian genetics*.

Genetic psychology The psychology of development (not of *genetics*). It covers the psychological development of both individuals and species but the term is no longer widely used.

Genital stage In *psychoanalytic* theory, the final stage of psychosexual development, beginning at puberty, in which the adult forms of sexual desire and activity are acquired.

Genotype The term used to describe the set of genes possessed by the individual, which have been replicated throughout the cells of the body. Because many genes will not encounter the circumstances in which they would become active, and others are *recessive* so that their action is suppressed by a *dominant gene*, less than 50% of the characteristics coded in the genes actively contribute to the structure or characteristics of the individual (the *phenotype*). However all of the components of the genotype are available for passing on to offspring.

Gestalt principles of perception An attempt to describe the important features of perceptual functioning through a set of principles which are consistent with the *gestalt* emphasis on wholes. See *law of Prägnanz*.

Gestalt psychology A form of psychology popular in Europe in the first half of the twentieth century, which gathered support in opposition to the mechanistic approach of the *behaviourist* school in America. Gestalt psychology emphasizes the holistic nature of the human being, and opposes *stimulus–response* reductionism, on the grounds that 'the whole is more than the sum of its parts', and that there are many aspects of perception, memory and learning processes which cannot be understood in terms of collections of smaller units, but which are complete and unitary in themselves. The Gestalt emphasis on *cognitive psychology* provided an important background to the 'cognitive revolution' of the 1960s and 1970s.

Gestalt therapy A method of *psychotherapy* developed by Fritz Perls which works in the 'here-and-now' rather than the past, and aims to increase the person's awareness of how all of their psychological processes are integrated. The emphasis on understanding the person as a whole is derived from *gestalt* principles.

Gesture A mode of *non-verbal communication* in which information is conveyed by movement, usually (but not always) of the hands and arms. Gestures tend to vary considerably from one culture to another; and the same sign may have a very different meaning even in neighbouring countries. [f.]

GIGO An acronym for 'garbage in garbage out' which was produced by computer folk to point out that if your data are rubbish, then running them through a sophisticated statistical programme on a computer will just turn them into different rubbish.

Glial cells Small cells which are found among the *neurones* of the nervous system. Their main function seems to be to provide nutrients and to absorb excreted waste from the neurones themselves.

Global attributions *Attributions* in which the chosen cause is of kind that is likely to affect significant outcomes. Such as, for example, attributing your unusually intensive revision to the fact that the result of the exams that you are taking will influence your whole future career. When the cause has only minor consequences it is called specific.

Gonads General term used to describe the sex glands; either the *testes* or the ovaries.

GPS See *general problem-solver*.

Gradient of colour One of the *monocular depth cues*, which indicates the distance of objects from the observer. Gradient of colour refers to the way that the colours of distant objects appear to be greyer and less vivid than those of near objects.

The use of gesture in communication

Accordingly, the brain utilizes the relative intensity of the colour to deduce probable distances.

Gradient of texture One of the *monocular depth cues* which indicates the distance of objects from the observer by utilizing the extent to which fine details of texture can be discriminated. Nearer objects present a finely detailed appearance, whereas those which are farther away appear to be 'smoothed out', and detail is lost. [f.]

Grammar A set of rules set up to attempt to specify how a language is constructed. Grammar is more concerned with *syntax* than with *semantics*, but particularly within *psycholinguistics* is likely to be concerned with both. The objective of a grammar is to have a set of rules which will generate all acceptable sentences within a language but no others. As with *logic* there may be a problem that a particular set of rules that does the job may not correspond to the rules that humans use to achieve the same objective. The most widely accepted form of grammar in psycholinguistics is that produced by Noam Chomsky and called *transformational grammar*.

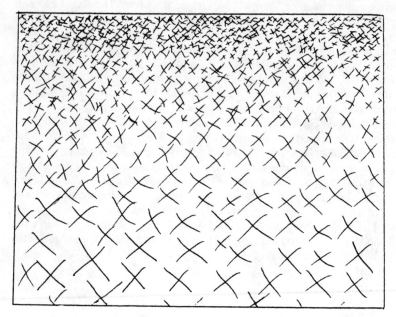

Gradient of texture

Group polarization The *emergent property* of groups in which people can be shown
to make more extreme decisions when acting as a group than when they are acting
as individuals. This was first identified with the *risky shift phenomenon*, in which
groups were shown to make riskier decisions than would be made by the individuals
which comprise them. However, more recent research shows that under certain
conditions groups may also make more cautious decisions, so the term group
polarization was adopted as one which described the phenomenon without making
assumptions as to the direction in which the group would shift.

Group test A psychometric test which is administered to several people at once, by a
single tester; such as some of the school-type *intelligence tests*.

Group therapy *Psychotherapy* carried out with a number of people, who come
together at agreed intervals for the purposes of the therapy. There are many
different forms of group therapy, corresponding to most of the varieties of
individual therapy, but all could be claimed to have two major advantages over
individual therapy. First, cost-effectiveness, as a single therapist sees several
patients simultaneously. Second, group processes ensure that the participants will
have genuine interactions with each other and real emotional experiences which
will relate meaningfully to their experiences in daily life, and which can be used
effectively by the group therapist.

Groupthink One of the *emergent properties* which can occur in tightly knit groups
with a high level of consensus and group loyalty. Groupthink is the phenomenon
in which a consensual view of reality emerges within the group, so that unpleasant
(and more realistic) alternatives to the way that the group sees the world are not
taken into consideration. Indeed, attempts to bring more realistic perspectives to
bear on the situation are often perceived as being disloyal. The process therefore

results in the group making silly, or at times tragic, decisions. The classic example of groupthink was the American military decision to invade Cuba at the Bay of Pigs – a decision which turned out to be a military disaster, and one which could have been easily foreseen, if those making the decision had been able to make a realistic appraisal of the situation. Groupthink can occur in any tightly knit group with strong leadership, and was therefore also apparent in some of the governmental decisions made during the Thatcher years in Britain. Investigations of groupthink suggest that conscious and deliberate efforts to promote debate and to admit unwelcome possibilities are required to overcome it.

Growth motive A term used in *humanistic* models of *personality* to describe the tendency of human beings towards personal growth and development, not only through the acquisition of new skills and experience, but also through cognitive re-evaluation and an increased sense of personal control and autonomy. Humanistic psychologists consider this to be a very basic motive in the human being, and fundamental to an understanding of mentally healthy behaviour.

Grey matter The term given to the densely packed mass of neuronal cell bodies and unmyelinated fibres found on the inside of the spinal cord and on the outside of the cerebrum. See also *white matter*.

GSR See *galvanic skin response*.

H

Habit In behaviourist terms, a habit is described simply as a learned *stimulus–response* sequence; in cognitive psychology it is seen as a set of automatic routines and sub-routines in which the individual engages, and which, owing to frequent exercise, requires little conscious cognitive input. The learning process involved in acquiring a habit is likely to involve *classical conditioning*, but will not be *habituation*.

Habituation A very basic form of learning which involves gradually ceasing to respond to a non-significant stimulus which is repeatedly experienced. Ceasing to notice the ticking of a clock is a typical example. Habituation can be distinguished from fatigue by the fact that a small change in the stimulus will result in the response reappearing, a process called 'dishabituation'. Habituation is essential in allowing organisms to concentrate on those properties of stimuli which have significance for them, and to avoid having the cognitive system overloaded with irrelevant information. So, for example, car drivers do not habituate to the sight of red at the top of a traffic light, but they are likely to have difficulty in remembering the colour of the stripes the poles are painted.

Haemophilia A genetic disorder which results in excessive bleeding when even slightly wounded, owing to an inability of the blood to clot. As a classic example of a *sex-linked trait*, haemophilia is found in many psychology text-books, although the psychological implications of the disorder are obscure.

Hallucination A vivid and convincing mental image which may appear in any sensory modality. The person experiencing it may be unable to believe that no sensory stimulation was involved. Hallucinations are taken as one of the most reliable signs of *schizophrenia* though they may be caused as a side effect of *psychoactive drugs.*

Hallucinogen A drug which induces *hallucinations* or other unusual forms of perception. The most commonly used hallucinogens are psilocybin and LSD (*lysergic acid diethylamide*), but there are many others, including mescalin and that contained in the fly agaric mushroom. Traditionally, hallucinogens have formed an integral part of religious and social ceremonies in many parts of the world. In the West they are normally used as *recreational drugs*, although there have been several instances of artists and creative writers utilizing their effects to obtain special insights for their work, and one or two investigations of their usefulness in certain kinds of *therapy*.

Halo effect An effect in which people or objects who are judged positively on one characteristic are also judged positively on others. For instance, a person who is judged to be physically attractive is more likely to also be perceived as being more amusing, or intelligent, than a physically less attractive individual of similar personality.

Handedness The term for specialization in use of one hand which develops in humans during the first years of life. Often the preferred foot or the preferred eye are not on the same side as the preferred hand. Handedness is thought by some to be related to *hemisphere dominance*. Since the right cerebral hemisphere controls the left side of the body and vice versa, people who are right-handed are thought to be left-hemisphere dominant, while left-handed people are right-hemisphere dominant. The evidence relating handedness to cerebral dominance is at times contradictory, despite the plausibility of the idea.

Hawthorne effect The phenomenon that when changes are introduced into a work environment in order to bring about an increase in productivity, there may be a temporary increase in productivity just because changes have been tried. An entirely useless change may therefore appear to work unless the effects are tested over a reasonable period. Hawthorne effects illustrate the importance of social factors and expectations in the working environment.

Hedonic relevance The issue of whether a cause leads to effects that have direct positive or negative consequences for the person concerned. A cause has hedonic relevance for someone if it produces something pleasant or unpleasant – so, for instance, a government ruling that student income was to be halved would have direct hedonic relevance for students. It would not, however, be personalized. See *personalism*.

Hedonism In philosophy, hedonism is the idea that pleasure or happiness is the highest good. In psychology, it is the idea that it is fundamental to human beings to seek pleasure and to avoid pain, and that this in itself is a valid explanation of much behaviour.

Helplessness theory An approach to human functioning deriving from Seligman's studies of *learned helplessness* in animals. Some animals were found to react to unpleasant situations over which they had no control by ceasing all attempts to change the situation. Their state of passivity and apathy was felt to resemble depression and so a theory that depression results from experiences of helplessness was proposed by Seligman in the mid 1970s. Subsequently the theory has been revised and integrated with *attribution theory*.

Hemispheric dominance The observation that, in most individuals, one side of the brain is more influential or has greater control over the body than the other side, thus possibly producing right or left *handedness*, etc.

Hemispherectomy An operation involving the removal of one entire *cerebral hemisphere*. Studies of left hemispherectomy in severely brain-damaged patients have shown interesting, often puzzling, recovery of language functioning and linguistic memory which was not evident when the damaged hemisphere was in situ. Such cases call into question the accepted idea that language is firmly localized on the left hemisphere, but rather suggest a hologram-like storage mechanism, whereby each hemisphere is capable of taking over the functions of the other, but does not do so in everyday functioning.

Heredity The processes by which part of the biological potential of the parent is transmitted to the offspring. In sexual reproduction this involves half of the genetic material of each parent combining to form the complete genetic structure of the offspring. See also *chromosome, gene*.

Heritability estimate A figure which purports to state the proportion of influence exerted by *genes* on the individual's development. Despite the fact that many developmental geneticists and psychologists (e.g. Hebb) have demonstrated unequivocally how inseparable genetics and the environment are, such figures continue to be constructed. The most well-known 'heritability estimate' is that of 80% genetic influence on the variation in intelligence, put forward by Jensen in 1969 on the basis of Cyril Burt's fraudulent data on twin studies. Controversy concerns not so much the estimate of 80% as the conclusions to be drawn from any estimate of heritability.

Hermaphrodite An individual who possesses the primary sexual characteristics of both sexes at the same time. True hermaphrodites have gonads, one of which has developed as an ovary and the other as a testis. They can therefore produce an ovum and fertilize it themselves and so potentially produce offspring without assistance from any other person. The condition is extremely rare and is not likely to be the true explanation of unexpected pregnancies.

Hermeneutics The study of meanings in social behaviour and experience. It is concerned with meanings on a number of levels, which range through the conscious and unconscious, personal, and social to the cultural and socio-political levels. Rather than simply looking at the generalities of behaviour, or at statistical information, hermeneutics is concerned with the interpretation of experience, and the ways in which various forms of symbolism are used to convey meaning in human life.

Heroin Heroin is a powerful *analgesic* of the opiate group, originally developed as a non-addictive pain-killer. However, it was soon found that as a substance it is extremely *addictive*, producing *tolerance* very rapidly and leading to increased doses of the drug being necessary for the same effect. It is probably the most abused of all the narcotic drugs. In addition to its analgesic properties, heroin induces profound mood-changes, leading to relief from tension and producing a state of drowsy contentment. Accordingly, its use and abuse as an illegal drug is most widespread in the poorer sector, but occurs to some extent throughout society. Addiction to heroin, in addition to the problem of tolerance, produces high susceptibility to infection and disease. In chemical terms, heroin has a structure very similar to the *endorphins* and *enkephalins* produced naturally in the brain in response to prolonged exercise, and is picked up at the same receptor sites.

Hertz A measure of frequency, one hertz being one cycle per second. In the audible range the frequency determines the pitch of a sound or tone. Tones of higher pitch produce more frequent cycles, hence they are said to be of a higher frequency.

Heterogeneity Varied, or showing a large number of differences. A heterogeneous sample is one in which the subjects are of many different kinds. Hetero- as a prefix means 'different' or 'other'. See also *homogeneity*.

Heteronomous morality The second of Piaget's stages of *moral development*, this is also known as the 'moral realism' stage. At this point, morality is considered to be subject to the laws of others; in other words, the child accepts as right and proper the rules given by authority.

Heterosexual Having sexual inclinations towards members of the other sex. See also *homosexual*.

Heuristic Problem-solving strategies which involve taking the most probable or likely options from a possible set, rather than working systematically through all possible alternatives. Heuristics provide a way of reducing a complex problem to a manageable set of tasks with only a slight risk that the solution lies among the alternatives excluded at the start. Heuristics differ from *algorithms* in that they do not guarantee a solution. See *problem-solving*.

Hidden observer The term given to the experience of a dispassionate 'inner self' which observes the individual in stressful situations, or during day to day living. Such an experience is particularly common during hypnosis, in which the hidden observer is felt to have experiences which are parallel to, but not the same as, the hypnotized self. In psychotherapy, the objective part of the therapist which comments on his or her feelings and involvement with the patient is called the 'observing ego'.

Hierarchy A structured form of organization constructed in levels, with each level overshadowing or dominating the lower ones. The idea of hierarchy is used in many different ways: a hierarchy of concepts, for instance, refers to the ways in which concepts may be stored in the brain, such that general concepts contain within themselves smaller constituent units. The analysis of organizations is almost always formulated in terms of hierarchies.

Hierarchy of needs Maslow's hierarchy of human needs refers to the idea that needs become important in a systematic progression. Lower, more 'basic' needs such as food and security are important first, and 'higher' needs such as for beauty and *self-actualization* only become important once the lower levels have been satisfied. The theory applies both developmentally and to the mature person. According to Maslow children must be adequately satisfied at one level before they start to develop motivations at the next level, so the higher stages are not reached for several years and self-actualization may take at least 30 years to achieve. Adults may be stuck at a low level if they have never experienced sufficient satisfaction at that level, but even those who have progressed higher may cease to be motivated at the upper levels if they are seriously threatened in a more basic way. For example, the need for dignity ceases to matter if you look up and find you are in danger of being run down by a bus. [f.]

Higher order conditioning See *secondary conditioning*.

Hippocampus A part of the *limbic system*, which seems to be particularly concerned with memory processing, in that people with surgical damage to both sides of the hippocampus have subsequently experienced an inability to store or recall new information, although earlier memories remain intact and can be retrieved at will.

Holistic Complete, treating its subject matter as a coherent and indivisible unit. For

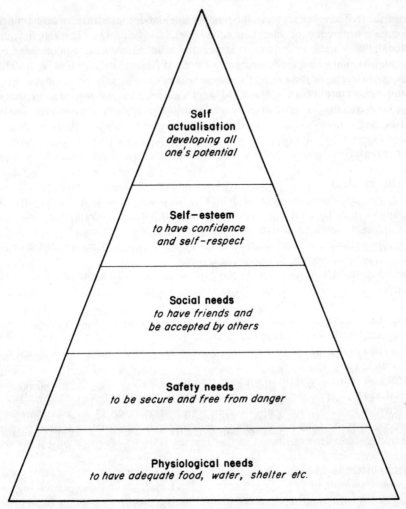

Maslow's hierarchy of human needs

instance, a holistic approach to medicine would involve dealing with the whole person, including their own experiences, stresses, and understanding of the situation, rather than simply treating the symptoms.

Hologram A portrayal of a three-dimensional image as a projection from a small unit or a two-dimensional surface, in such a way that the image can be examined from different angles and shows the appropriate parallax. In addition, a complete holographic image may be reconstructed from a proportion of an original, although some clarity of detail is lost. Understanding the perception of holograms poses a unique problem to psychology, which as yet seems far from resolution. Holograms are also of interest to psychologists because in some ways the *association cortex* functions similarly; it seems likely that information is not stored in a specific location but is available in any large enough area of the cortex.

Holophrase A single-word utterance which conveys the meaning of a whole sentence in itself, e.g. 'beaten!'.

Homeostasis The process of maintaining a stable condition or state by detecting and reducing differences from a goal state. The classic simple example is a central heating system where the thermostat turns the boiler on when the temperature drops and off when the temperature is high enough. The basic process involved is called *negative feedback*. The concept has been widely used to describe the maintenance of physiological balance in the body, with metabolic functions kept at an optimal level through the operation of mechanisms which correct imbalances. Homeostasis in the human body is maintained through a variety of mechanisms, tightly mediated by the *hypothalamus*. *Drives* are considered to arise directly from such homeostatic mechanisms: the hunger drive, for instance, is purportedly initiated when blood-sugar levels in the body fall below a certain level. This produces food-seeking behaviour, until food is ingested and satiation is reached. The concept of homeostasis plays an important part in systems theory and can therefore be applied to how psychological stability is maintained in people and their families.

Homogeneity Similarity or likeness. Something which is homogeneous is the same overall, showing little variability. A homogeneous group of subjects will have been selected so that all of them score similarly on essential measures. For example one might recruit a sample of 25-year-old middle-class mothers, each with one preschool child. This would be a homogeneous sample for research on child rearing (though not necessarily for research in other fields, like religious attitudes). A mixed sample is called *heterogeneous*. The prefix homo- means 'the same'. It is not related to the Latin 'homo' meaning 'man'.

Homogeneity of variance One of the criteria used for the selection of a *parametric test*. Homogeneity of variance refers to the variance, or 'spread', shown by the populations from which the data samples have been taken. The purpose of parametric analytical techniques like the *t-test* is to compare the *means* of two samples, in order to see if they are different enough to have come from different populations. But the formula for estimating the variance of the parent population

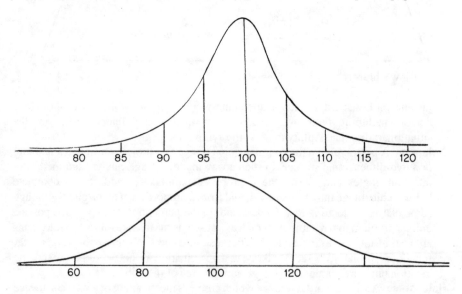

Two normal distribution curves with differing variances

relies on the two sample *variances* being similar, so if they are not, a t-value would be misleading. For this reason, homogeneity of variance is an important criterion for using a t-test; and variance is usually checked using an F-test. Note that finding a significant difference in the variances of the two samples might be just as important as finding a difference in their means. [f.]

Homophobia Hostility to homosexuals as a group. See *reaction formation*.

Homosexual Having sexual inclinations towards others of the same sex. The prefix homo- is from the Greek meaning 'the same'; and not the Latin meaning 'man'. The term therefore applies equally to men and women. See also *heterosexual*.

Hormones Chemicals released into the bloodstream which produce changes in the functioning of the body. Hormones are produced by the glands of the *endocrine system*, which operate in close conjunction with the *hypothalamus*.

Hostile aggression Aggression in which the objective is to inflict harm on the other, as opposed to *instrumental aggression* which is undertaken for some other purpose. See *aggression*.

Hue The term used to describe a particular wavelength, or tint, of a colour. It is a subdivision of the broader categories of colours; so, for instance, we have different hues of green.

Humanistic psychology An approach within psychology which emphasizes the whole person and their scope for change. Humanistic psychologists reject the *reductionist* approach of many researchers, which sees human action simply as collections of separate mechanisms; and they also argue against the dehumanization and 'objectifying' of human behaviour produced by trivial laboratory investigations and *behaviouristic* attitudes within psychology. Instead, they argue that psychologists should take more account of the whole person, including *attitudes*, values and responses to social situations (including experiments). To attempt to study people in a fragmented way is, they think, to ignore the essence of what it is to be human. There are many humanistic psychologists, of whom Carl Rogers is perhaps the most famous. Humanistic psychology is also closely linked with the *phenomenological* approach within psychology.

Hygiene factors Factors in the working environment, identified by Hertzberg, which are to do with the working conditions of the individual, such as shift organization, staff facilities, and organizational structure. In investigations of job satisfaction, Hertzberg found that bad hygiene factors contributed considerably to job dissatisfaction, but that incentives known as 'motivators' (e.g. promotion prospects, a sense of goals, etc.) were necessary to produce job satisfaction itself.

Hyper- A prefix indicating a high or excessive level of some function.

Hyperactivity or hyperkinesis A condition of excessive and apparently uncontrollable activity in children. There is controversy over the reality of the condition, but there are some children whose activity is maintained at such an extreme level that the label seems to be unavoidable. It is also clear that many children who are labelled as hyperactive are just rather more active than their parents or teachers find convenient. The condition is strongly associated with difficulties in maintaining attention, leading to boredom, and it may be this aspect, rather than the activity level itself, which is fundamental. Hyperactivity can be effectively treated with drugs related to *amphetamines*. Although these drugs are usually used as stimulants, they also help in maintaining attention, and it seems to be this effect that is useful to hyperactive children.

Hypercomplex cell A type of cell discovered by Hubel and Wiesel, located in the

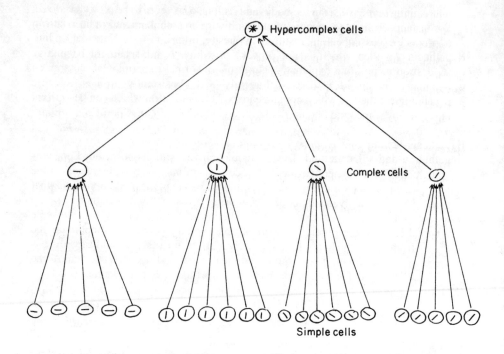

The receptive fields of simple, complex and hypercomplex cells

thalamus and the *visual cortex*, which receives information from *complex cells* concerning basic images occurring in the visual field. Hypercomplex cells collate this information, so as to respond to simple patterns and shapes, and are thought to represent the basis of *figure–ground perception*. [f.]

Hyperphagia Excessive eating which may be induced by lesions to the hypothalamus. See also *set-weight*.

Hyperlexic Hyperlexic children are those who learn to read extremely quickly, with little apparent difficulty. See also *dyslexic*.

Hypnagogic imagery Vivid visual imagery which is experienced during the transition from waking to sleep. It often takes the form of an unusually clear image of an object that has been the subject of intense concentration during the day, but the most common image is of falling. Hypnopompic imagery, which is rarer, is a similar kind of imagery which occurs during waking.

Hypnosis An *altered state of consciousness*, usually induced by voluntarily allowing one's actions to be directed by another person (the hypnotist). The major characteristics of the state are heightened suggestibility and concentration of attention on the hypnotist. Some subjects appear to achieve a very high level of this state, called a hypnotic *trance*, in which they are totally controlled by the hypnotist. Some have argued that hypnosis is just a matter of suggestible people role-playing a trance state, although others have found evidence of EEG records of a changed pattern of brain activity during hypnotized states.

Hypothalamus A small but important part of the brain, located immediately below the thalamus (hence the name). The hypothalamus is generally concerned with maintaining *homeostasis* in the body; and its functions seem to be partly localized. For instance, lesions to specific nuclei within the hypothalamus have been shown to produce excessive eating, resulting in obesity, in rats.

Hypothesis An idea which is not proven, or which is advanced as a tentative suggestion or possible explanation. In terms of formal experimental method, an hypothesis is an idea, derived logically and consistently from a specific psychological theory, which contains an explicit prediction which can be verified or refuted by some kind of empirical investigation, usually an experiment. See *null hypothesis.*

Hypothesis-testing See *hypothetico-deductive method.*

Hypothetico-deductive method The technique of investigation outlined by Popper as being central to the scientific method. It consists of investigating by means of the formulation of an explicit *hypothesis* containing an explicit prediction as to what would happen in a given situation. An empirical investigation would then be set up to test the hypothesis, i.e, to see if the prediction were true. If the hypothesis were retained because the prediction worked, that would be taken as support for the theory from which the hypothesis was derived. If, on the other hand, the hypothesis were refuted, that would be taken (in an idealized world) as evidence against the original theory; and an alternative explanation would have to be found.

Hysteria A physical symptom, with no apparent physical cause, but which appears to have some psychological function. An example would be temporary blindness or a paralysed arm which prevented someone performing a job which they hated but dared not leave. Such symptoms are not under voluntary control.

Hz See *hertz.*

I

Iconic representation The coding or representing of memories by utilizing sensory images (from the Greek 'icon' meaning 'image'). Iconic representation is usually used to refer to visual imagery, and was considered to be the second *mode of representation* to develop, according to Bruner. See also *enactive representation, symbolic representation.*

Id The primitive part of the unconscious personality, according to Freud, characterized by extreme emotional reactions and demands for immediate gratification. The function of the id is to fulfil the instinctual needs but it operates according to the *pleasure principle* and may be satisfied by fantasizing the desired object. Therefore the infant has to begin to develop the *ego* to deal with reality.

Ideal self-image The internalized concept of the perfect version of ourselves, which, according to Rogers, every individual has. The ideal self-concept is used as a yardstick by which the actual self's behaviour is judged: accordingly, it expresses the individual's internalized *conditions of worth*. Highly *neurotic* clients are often distinguished by an unrealistically high ideal *self-concept*, resulting in continual anxiety and a recurrent sense of failure.

Identical twins See *monozygotic twins.*

Identification A process seen as essential by both social learning and psychoanalytic theorists because it is an efficient way of acquiring new characteristics. It is the second stage of the social-learning process outlined by Bandura; the first of which is *imitation*. Identification refers to the internalization of imitative learning, such that it becomes incorporated into the individual's *self-concept*. So, for instance, a person starting a new job may spend the first couple of days consciously imitating others in that role. After a while, they come to internalize the new role, and are able to generalize their learning to novel situations. Freud proposed that models would be chosen when they were seen as successful in solving those problems which the person found most urgent, or had power over them. During the *oedipal* phase the strongest identification is with the parent of the same sex as the child, and so an appropriate gender identity is formed. A person may identify with a particular individual, or with a particular social role.

Identification figure See *role model*.

Identity The sense an individual has of the kind of person that they are. According to Eric Erikson the major task of adolescence is to establish a stable sense of identity which will remain relatively constant as the person moves between different situations. A failure to achieve a secure identity results in identity diffusion, which leaves the young adult unable to enter into commitments or close relationships for fear of being taken over by the other person.

Identity formation The process of forming an identity. The identifications made throughout development play an important role, and adolescents in particular will try out different kinds of identity and use feedback from others to decide which to retain and which to abandon.

Idiographic Attempting to understand the functioning of individuals, as opposed to the search for general laws of behaviour. Idiographic approaches to human *personality* examine characteristics which are considered to be common to all individuals, but which, in their operation, make each person unique. So, for instance, *personal construct theory* represents an idiographic approach; whereas the *psychometric* approaches, which are concerned with comparing people with one another, do not. See also *nomothetic*.

Idiosyncratic Special to that particular individual; characteristic of that person but not of most people.

Idiot savants People of very low general intelligence who have an exceptional ability in one specific area, such as being able to do very elaborate mental arithmetic extremely quickly.

Illusion Something which tricks the senses into a false interpretation of what is there. Illusions may operate in any sensory mode, but the most well-understood ones are *visual illusions*. These have been extensively studied because they offer a chance to see how the visual system works. Many visual illusions seem to have their effect by mobilizing distance constancy mechanisms.

Illustrators *Non-verbal* signals which serve to amplify or to demonstrate what someone is saying. See also *affect displays, emblems*.

Imagery Mental representations which recreate sensory impressions: visual imagery refers to an impression of something as it would be directly seen; auditory imagery is a representation of something being heard. An image is usually of a fairly specific object, but may sometimes be more diffuse, e.g. of autumnal colours. The study of imagery has been a major area in memory research, as it forms one of the main systems for the *encoding* and representation of memories. See also *hallucination, iconic representation*.

Imitation The copying of a specific action or sequence of behaviour. Imitation forms a learning process which is very common among all mammals, and especially humans. It provides an extremely rapid form of learning and a mechanism of early socialization. See also *identification*.

Immediacy of reinforcement The concept in *operant conditioning* that, in order for a particular behaviour to be learned, it must be reinforced immediately – i.e. as soon as it has taken place. Delayed reinforcements could mean that alternative behaviours occur in the meantime, and become accidentally strengthened through becoming associated with the reinforcement. See also *law of effect*.

Immediate memory A term occasionally used instead of *short-term memory*.

Implicit personality theory The ideas about how *personality* traits are grouped together often taken for granted in everyday living. For example, traits like ambitious may automatically be grouped with aggressive and energetic; or kind could be grouped with gentle and peaceable. This means that individuals who are known to have one particular characteristic are often reacted towards as if they also possessed the full range of associated traits. They are treated in accordance with the unspoken and assumed theory of personality held by the people whom they encounter. See also *personal construct theory*.

Implosion therapy Otherwise known as flooding, this refers to a technique in *behaviour therapy* in which the *phobic* individual receives direct and extended exposure to the feared stimulus, until they become relaxed with it. For instance, someone who has had a car accident and become frightened of going out may be repeatedly shown film of cars approaching them. As they become used to this, the fear dies away and, through *classical conditioning*, a more relaxed attitude becomes associated with the stimulus. See also *systematic desensitization*.

Imprinting The process involving following of the mother which occurs during a *critical period* shortly after birth in some species, most notably in ducks and geese. The 'following' behaviour can be elicited by any moving object during the hours, or perhaps days, after birth, and the animal seems to have a strong *innate* tendency to learn about and in some way identify with the object. The learning is very resistant to change, and later in life social and sexual behaviour will be directed at animals or objects which resemble the imprinted stimulus. Attempts have been made to explain the *attachment* of human infants to their mothers as a form of imprinting but the two processes are quite different, and it seems that the main thing they had in common at the time the theory was proposed was that neither could be satisfactorily explained. [f.]

Incentive A stimulus which has value, either positive or negative, for an organism.

Incentive theory A theory of motivation that distinguishes between the expectation that a goal can be achieved (incentive motivation) and the strength of the need for the goal (*drive motivation*). The amount of effort made to achieve a goal is a function of both kinds of motivation. So high drive alone may be ineffective if paired with low incentive: I would very much like a million pounds but do not expect success so I am not doing anything about it. Equally, high incentive (I am sure I could get spam for dinner if I tried) will not generate goal (or spam) seeking if my drive is low because I do not like the stuff. Practically, the theory indicates that if an organism is not working for a goal it is necessary to know whether to increase need (life will be really wonderful if I can get A-level psychology), or incentive (there is still enough time to look up all the terms I do not understand).

Incubation period The period during the creative process in which ideas seem to

Imprinting

develop and become formulated at a totally subconscious level. Typically this has been preceded by an acquisition period, in which ideas are experimented with and tested out and is followed by a period of insight, and then intense creative activity, in which the artist-writer-creator produces the final work. Although not all creative individuals appear to operate within this four-stage model, it seems to be a common sequence for many; and the incubation period, in which work on the idea seems, on the surface, to have ceased, is its distinctive feature. See *creativity*.

Independent-measures design The kind of study that involves comparing the scores or responses from two or more separate groups of people, such that one group experiences one of the experimental conditions and the other group experiences a different condition. See *repeated-measures design*.

Independent variable The variable that an experimenter sets up to cause an effect in an experiment. An independent variable may have two or more conditions, and subjects' responses to each of them are studied. Independent variables may be existing features of the subjects (males vs females) or be created by the experiment (dark vs light conditions). It is called independent because it is not affected by the experimental procedures. See *dependent variable*.

Individual differences The study and measurement of the significant ways in which individuals differ from each other. Some studies of *individual differences* deal only with intelligence test scores, but the area is usually taken to include any reasonably stable characteristics or abilities. It therefore includes personality traits and psychological dysfunctions.

Individuation The process of becoming separate. It is used particularly about people during the transition from adolescence to adulthood when they separate from and become independent of their families. Jung felt that individuation could not be fully achieved before middle age.

Induction Making general laws from knowledge of particular cases. Inductive reasoning is being used when results from a sample are utilized to make statements about a population, and is fundamental to the operation of empirical psychology. Inductive reasoning has also been studied as part of the subject matter of cognitive psychology. It can be contrasted with deduction.

Industrial psychology The application of psychology to industrial situations. Industrial psychologists may study the effects of *environmental* influences on people at work; of *organizational* influences, such as the effects of different management structures or styles; of social relationships within an industrial setting; or of sources of *stress* and industrial accidents.

Infancy The period of human development before the child is able to speak. Usually taken as the first year or two of life.

Infantile autism See *autism*.

Infantile sexuality A supposition, originating with Freud, that the sensual pleasures and motivations of infants have a sexual basis. The issue became one of great controversy and in some respects rests on issues of the definition of sexuality. However, it is also the case that Freud was indicating a previously unrecognized aspect of infant functioning when he pointed out the pleasure that all infants obtain from activities like oral stimulation and masturbation.

Information-processing An approach which analyses cognitive processes in terms of the manipulations of information that are involved. As computers have become capable of progressively more sophisticated operations, information processing has become accepted as a plausible approach to understanding *perception*, decision making, etc. The approach is more directly involved with computers when they are used to run models of particular cognitive processes (called a *simulation*) to see how the model would work in practice.

Inhibition (i) The process by which a *neurone* becomes less likely to fire. Inhibitory synapses are those which produce a raising of the *threshold of response* for the next neurone, thus rendering it likely to fire only in response to extreme stimuli.

(ii) A process in learning by which a response becomes increasingly less likely with repeated presentations of the stimulus. The term inhibition is generally used to refer to a damping down or restraining of a behaviour, as a result of over-use or some other kind of direct stimulation.

(iii) The idea of a specific memory becoming lost or distorted as a result of further information. See *interference*.

Inhibitory synapse A synapse which operates in such a way that the nerve cell which receives its message becomes less rather than more, likely to fire so the passage of the neural message is inhibited, rather than passed on. Compare *excitatory synapse*.

Innate Literally, inborn. It means unlearned, or present at birth, and is used synonymously with inherited or *genetic*. Compare *congenital*.

Innate releasing mechanism (IRM) A term used by Tinbergen to refer to the stimulus which triggered off an instinctive behaviour. Examples are the moving shape which stimulates pecking in a young herring-gull chick and that which provokes 'freezing' in turkey chicks. The behaviour released by an IRM has direct survival value, either in avoidance of predators or in obtaining food. Currently the term *sign stimulus* is preferred to refer to these signals, as it avoids the implicit assumptions about internal mechanisms contained within the term IRM.

Inner ear The third main division of the ear. It is that part of the ear with direct connections to the brain, via the auditory nerve. The inner ear contains the *cochlea*: a long, fluid-filled tube containing hair cells which transduce the vibrations of sound information into electrical impulses. It also contains the semicircular canals, which detect orientation of the body and motion, in a similar fashion (i.e. by hair cells which fire when stimulated by motion or vibration). See also *middle ear*.

Insight (i) In learning or creativity, a sudden and complete realization of the solution to a problem, usually involving a restructuring of the subject's perceptions. The process was regarded as particularly important by the *Gestalt* psychologists.

(ii) An awareness of one's own psychological processes, unconscious fears and wishes, etc. Forms of psychotherapy which work specifically to increase insight, such as *psychoanalytic* and *humanistic*, are called the insight therapies.

Instinct A term now avoided as much as possible, but once used to refer to those aspects of human experience which were deemed to have been inherited and to be immutable. The concept of an instinct is as being always directed towards function, such as 'for' security, motherhood, etc. and is therefore of very little value in describing or explaining behaviour itself. An instinct for security might manifest itself in a variety of ways. To one person it might mean having money safely invested, to another it might mean having a comfortable home, to yet another it might mean becoming increasingly self-reliant and able to survive with as little money as possible. Such potential diversity of behaviour means that the concept itself is of dubious value, and has largely been replaced by the term *instinctive behaviour*.

Instinctive behaviour Behaviour which occurs as a result of the direct action of genes. Such behaviour typically shows certain distinctive characteristics. These are: *stereotype* (the behaviour being fixed and not modifiable by the individual); there is a complex sequence of behaviour, not just a reflex response; it arises in individuals even if reared apart from their own species; it does not require prior learning or practice; it is *species-specific*. Such behaviours would appear to be relatively common in fish and birds, but rather less so among the higher animals.

Institutionalization The effect on a person of living for a long time in an institution. Institutions like mental hospitals are likely to develop procedures which are very different from those in the outside world. As the inmate adapts to the regime they may develop patterns of motivation and behaviour which would prevent them from functioning successfully in the outside world. Ironically, the phenomenon operates most clearly in just those institutions, like mental hospitals and prisons, that are supposed to improve the client's ability to function within society. It has been suggested that the reason that staff in institutions fail to take the process of institutionalization into account is that they themselves are subject to it.

Instrumental aggression *Aggression* which occurs because it will result, directly or indirectly, in a desired outcome for the individual showing the aggression.

Instrumental learning Learning which occurs as a direct result of the beneficial or pleasant consequences which it has for that individual. Often used synonymously with *operant conditioning*.

Intellectualization A way of coping with anxieties by denying the emotional component of a situation and concentrating on an abstract logical account of the details of the situation and one's own response to it. One of the *defence mechanisms*.

Intelligence In general, the ability of an individual to understand the world and work out appropriate courses of action. Within psychology there is no more precise definition that is generally accepted, though the old claim that 'intelligence is what *intelligence tests* measure' is uncomfortably acute. See *intelligence A, B, C; intelligence quotient; intelligence test*. See also *triarchic intelligence*.

Intelligence A, B, C Classifications developed by Hebb and Vernon in an attempt to express the relative contributions of experience and inheritance to an individual's intelligence. The term Intelligence A was used to describe the total potential intelligence of an individual, given that particular genotype and an ideal environment from conception. Intelligence B was conceived as an unknown proportion of intelligence A, which was that amount of their potential which the individual had been able to realize throughout their life. Intelligence C referred to the unknown proportion of Intelligence B which would be measured using an *intelligence test*. In formulating this model, Hebb was applying the genetic distinction between *genotype* and *phenotype* and arguing that, to talk of the relative contributions of genetics and environment as if they were alternatives or could be quantified, was inherently misleading. [f.]

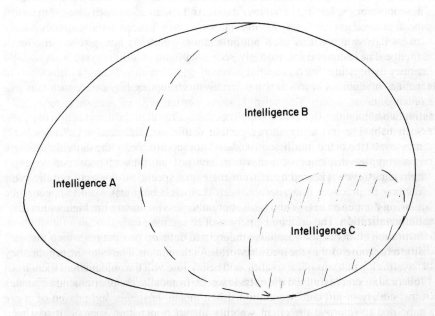

A possible model of intelligence A, B and C

Intelligence quotient (IQ) A score devised by Binet, as an attempt to express the relationship between a child's *mental age* and its actual, or *chronological age*. The quotient was obtained by dividing the child's mental age (obtained by using a variety of age-related tests) by its chronological age, and then multiplying the result by 100. This meant that 100 became the normative figure: a child which had a mental age appropriate for its chronological age would score 100; those who were advanced for their years would score above 100, and those who were behind would score below 100. Although Binet repeatedly expressed his concern that this should not be taken as indicative of a child's potential to learn, but simply of its achievements thus far, IQ scores have been systematically misused to represent a static measure of the individual's intellectual capacity. In addition, despite the normative nature of IQ scores, in many cases they have been erroneously treated as equal-interval data, and used as the basis of elaborate statistical calculations such as those underpinning the concept of *heritability*. Such research has formed the basis for such outcomes as compulsory sterilization laws in the United States (see *eugenics*), differential schooling systems in many countries, and has contributed to the concepts of racial inferiority which resulted in the attempted genocide of the Jews in the Second World War. Although the original formulation of IQ had considerable diagnostic value, its widespread misuse and abuse in society has resulted in its use being regarded with considerable suspicion.

Intelligence test A standardized set of tasks from which intelligence can be estimated. All tests should have been fully assessed for *reliability* and *validity*, but a great variety is now available, to some extent reflecting problems that have been identified during the history of mental testing. Of the most widely used tests, the *Stanford–Binet* is a direct descendant of the original test devised by Binet to give a single measure of IQ. The *Wechsler* provides 12 sub-scales measuring different aspects of intelligence. *Raven's Progressive Matrices* attempts to eliminate cultural bias by having items and even administration which do not depend on using language. The recently developed British Ability Scale is an attempt to incorporate later psychological work on intelligent performance, such as Piaget's ideas.

Interaction A situation in which one thing reciprocally affects another, such that an exchange takes place. The term is used particularly in reference to *social interaction*.

Interactionism The interactionist perspective within *physiological psychology* is a direct contrast to the traditional *reductionist* approaches. Rather than seeing physiology as the direct cause of behaviour, an interactionist perspective emphasizes the ways in which environment, cognition and physiology may all have a reciprocal effect on one another, such that each may influence the other in achieving a given effect. Within this approach, physiological variables which are usually regarded as causes may equally well be seen as results.

Interference The concept in *memory* theory that information may become lost or distorted because of the storage of additional information. The interference theory of *forgetting* was a popular approach in memory research throughout the 1950s and 1960s, and it centred round the idea that memories could become displaced because of the storage of similar information. Interference was considered to be of two kinds: *proactive interference*, in which material which had been learned first interfered with the acquisition of later information; and *retroactive interference*, in which information which had been acquired at a later stage interfered with the retrieval of previously learned material.

Intermittent reinforcement Reinforcement which is given only in some instances of the desired behaviour, and not every time the behaviour occurs. See *schedules of reinforcement*.

Internal attributions Attributions in which the chosen cause is internal to the person concerned, rather than arising from circumstances in which case they would be called external. For example, perceiving your exam success as having been caused by your own hard work and/or ability, rather than by luck. Internal causes are often equated to *dispositional attributions*, although this can sometimes produce conceptual difficulties, e.g. 'is hard work a disposition?'.

Internal–external scale A scale originally devised by Rotter in the 1950s to measure whether a person believes the causes of events to originate within themself (emotions, abilities, effort) or outside (powerful other people, luck). See *locus of control* for one use of such a scale, and *attribution theory* for another.

Internal validity The extent to which an individual item in a test measures the same thing as the other items. See *validity*.

Internalization Making something part of oneself. Freud was concerned with the child internalizing the moral values of its parents, as expressed in their system of rewards and punishments. The term is now used more broadly, particularly in areas like *conformity studies*, where its use distinguishes subjects who have fully adopted and internalized the values from those who express them for expediency.

Interneurones Neurones within the *central nervous system* which connect sensory input (brought by *sensory neurones*) with motor output (carried by *motor neurones*).

Interpersonal attraction The study of what determines whether a person will find another attractive. After decades of research investigating a great range of subtle variables it has emerged that people are attracted most to those that they find physically attractive and who are geographically close.

Interpretation The activity of making sense of information and identifying essential meanings. In *psychotherapy*, the activity of the therapist in pointing out underlying meanings in the patient's activities or cognitions. In *psychodynamic* therapy interpretations are made to uncover the *defence mechanisms* of the patient and to describe the patient's *transference* reactions, with the aim of making the patient's *unconscious* processes explicit.

Interval scale See *equal-interval scale*.

Intervening variable An unobservable process which is proposed to account for the relationship between input and output. The characteristics of intervening variables can be studied by manipulating the *independent variable* and observing the effects in the *dependent variable*.

Interview A conversation between a professional and a subject designed to provide the professional with a certain kind of information. The nature of the interview will be influenced by its function which may be evaluation of the subject (for a job), therapeutic, or research. The form of the interview may be fully specified in advance (structured interview), be planned in more or less detail, or be conducted without any prior consideration of what information is wanted and how it is to be obtained. Research has shown interviews to be an inaccurate method of selecting, but this may be because the interviews studied had not been carefully constructed.

Introspection The process of self-examination, or looking within one's own experience in order to gain insight into psychological phenomena. Although notoriously unreliable in many respects, introspection can at times provide valuable insights which could at other times be missed.

Introspectionism A school of thought, prevalent in the early years of psychology as an independent discipline from philosophy, in which investigations were conducted through systematic, and often detailed, introspection by one or two highly trained psychologists. Although often castigated as 'armchair psychology' by the early behaviourists, this technique established some important theoretical perspectives, such as those outlined in James's *Principles of Psychology*; which in many cases are still of use to modern psychology. With the advent of *behaviourism* in the first part of the twentieth century, introspectionism as a technique became disregarded; but of recent years it has re-emerged to a limited extent within the *phenomenological* school of modern psychologists.

Introvert An individual inclined towards a solitary, reflective lifestyle. Introversion is a personality dimension regarded as the opposite of *extraversion*. It was proposed by Carl Jung and incorporated into the *Eysenck Personality Inventory*. Eysenck sees introversion as rising from a higher level of cortical arousal resulting in a lower level of inhibition to the same stimulus. Introverted individuals, he argues, do not get bored as easily as extraverts, and so are better at tasks requiring sustained attention or which involve relatively little change in stimulation.

Involuntary response A reaction or reflex which is produced to a stimulus regardless of the individual's conscious intervention or inclinations. See also *unconditioned response*.

Ipsative Assessed or measured by comparison with the self. Ipsative scales involve the individual using his or her own values or behaviour as the yardstick by which comparisons and evaluations are made.

Ipsilateral Belonging to, or relevant to, the same side. The prefix 'ips-' usually means 'of one's own'.

IQ See *intelligence quotient*.

IRM See *innate releasing mechanism*.

J

James–Lange theory An early theory of emotion which argued that the experience of emotion arose from the perception of physiological changes in the body, brought about by the emotional stimulus. In other words: the physiological changes occurred first, and the emotion was simply the perception of those changes. See also *alarm reaction; Cannon–Bard theory*.

Jet lag A syndrome in which the individual's *circadian rhythms* become out of phase with the surrounding environment, as a result of the rapid crossing of time zones during long-distance travel. This produces feelings of extreme fatigue, and in some cases disorientation, sometimes lasting for several days until the individual adjusts fully to a new time system.

Jungian Pertaining to the psychoanalytic system developed by Carl Jung – sometimes also referred to as *analytical psychology*.

Just noticeable difference (jnd) The smallest change of stimulus which an individual is able to detect consistently 50% of the time. The amount of the jnd varies as a proportion of the intensity of the stimulus which is changing: for instance, a relatively larger change is necessary before a difference in volume of a loud sound is detected, than for a relatively quiet sound. See *Fechner's Law, Weber's Law*.

Juvenile delinquent A young person who has been convicted of a criminal offence.

K

Key word method A *mnemonic* technique for learning the meanings of technical or foreign terms, which involves identifying a key familiar word derived from the sound of the unknown one. By forming a visual image linking this key word with the meaning of the word to be learned, the information is acquired; the visual image forming a link between the perceived sound of the new word and its meaning.

Kibbutz An Israeli community in which property and responsibility are held in common by all members of the kibbutz (kibbutzniks). Many kibbutzim have communal child-rearing systems, which were intensively studied in the 1960s. The then current theoretical ideas on mother–infant *bonding* implied that children would become psychologically damaged if not kept with their mother. Little evidence of this was found among the communally reared children of the kibbutzim.

Kinaesthetic To do with sensations of movement. The kinaesthetic senses are those which are concerned with the detection of movements of the body: *proprioceptors*, for instance, detect balance, movements and orientation of the limbs.

Kinesics The study of human movement patterns and the types of communication which use them. Kinesics is a major area in the study of *non-verbal communication*, involving gestures and changes of posture and gait.

Kin selection A concept put forward in *sociobiology*, kin selection involves the idea that an individual may protect their genes for the future by protecting not just their offspring, but other relatives who share them. Since siblings share on average 50% of their genes, the individual can ensure that a proportion of the genes survive by protecting his or her siblings. The concept is used to explain behaviour which is apparently *altruistic*, such as the self-sacrificing behaviour of worker ants.

Klinefelter's syndrome A condition in which a man has inherited an extra X chromosome, having an XXY group of sex chromosomes instead of an XY pair. Such individuals are usually clearly male, but can sometimes show some female secondary sexual characteristics.

Korsakoff's syndrome A condition acquired by long-term *alcoholics* who have combined heavy drinking with eating too little, resulting in an extended period of thiamine deficiency. Korsakoff's syndrome patients demonstrate severe and apparently irreversible *proactive amnesia*, such that they are unable to retain new information, while still maintaining their repertoire of basic skills. While conversational topics remain on a general level, many Korsakoff sufferers remain undiagnosed. An attempt to retrieve current information often reveals the deficit.

L

Labelling When a label is applied to someone there is a tendency for that person to be seen, both by others and often by themselves as having all of the characteristics implied by the label, and being nothing more than that. So labelling someone as *schizophrenic* or *depressive* can cause them to be treated as less than a whole person, since all of their behaviour is likely to be interpreted in terms of the illness, as schizophrenic or depressed behaviour. This can be resisted by insisting on referring to 'a person with depression' rather than 'a depressive', but the tendency remains difficult to avoid. The study of labelling and its implications is an important part of *social psychology*, and has been so ever since the discovery of the *self-fulfilling prophecy*.

Labile Changeable, or likely to alter rapidly. Often used of emotional states or *autonomic* arousal.

LAD See *language acquisition device*.

Laissez-faire Leaving people to get on with things in their own way. It is used to indicate a style of *leadership* in which most of the responsibility for action is left with the group, rather than assumed by the leader. Groups with laissez-faire leadership tend not to be as productive as others, but some findings suggest that they continue to operate better than other groups when the leader is absent.

Lamarckian genetics The theory of genetic transmission proposed by Lamarck at the beginning of this century, this model proposed that characteristics which an individual acquires during its own lifetime can be passed on to its offspring. So, for example, giraffes had acquired long necks because they had had to stretch upwards for food, and the elongation caused by stretching had been inherited by the next generation. Although now thoroughly discredited as a model, Lamarckian genetics influenced a number of other theories, most notably Piaget's model of *cognitive development*. See also *Mendelian genetics, genetics*.

Language The complex system of *communication* which involves the organization of words into meaningful combinations. Although most people would agree that the use of language is a distinctively human attribute, the lack of a precise definition of what exactly language is, makes it difficult to decide whether such phenomena as bird songs, bee dances, or whatever can be taught to chimpanzees in this line, should be called language. It is generally accepted, however, that language involves symbolic representation, and that there are distinct rules concerning acceptable combinations of the elements of language (usually words) which do not permit all possible combinations to be regarded as meaningful.

Language can be studied on a number of levels, which may be broadly classified as lexical (concerning the word units themselves and their referents); *syntactic* (concerning the rules for combining words into meaningful utterances); and *semantic* concerning the meaning of what is said). The use of analogy and metaphor in language means that the lexical characteristics of an utterance may not be identical with its semantic characteristics (e.g. describing someone as 'burning' with enthusiasm). Psychologists have also studied social aspects of language use; such as the impact of accents or sexist language; and recently much research attention has been devoted to discourse analysis: looking at the way that language is used in complete conversations. See also *paralanguage, psycholinguistics, sociolinguistics, verbal deprivation.*

Language acquisition device A mechanism proposed by Chomsky to explain the entreme rapidity with which young children develop speech. He proposed that the young infant is born with an innate language acquisition device, which enables it to extract basic rules of grammar from the speech heard around them. Moreover, this occurs as a more-or-less automatic process: all that is required is that the child hears or experiences language used by others. In view of an increasing body of research indicating that human interaction forms a fundamental part of speech acquisition, later theorists have modified this concept, preferring instead to talk of a language acquisition system, or LAS, which allows for rather more active involvement on the part of the child than simply passive decoding. See also *deep structure, surface structure.*

Language areas Specific parts of the cerebral cortex, usually (though not always) located on the left hemisphere, and mediating the functions of language. There are three main language areas: *Broca's area*, which is largely responsible for speech production and the formulation of appropriate words, *Wernicke's area*, which is concerned with the comprehension of speech, and the *angular gyrus*, which receives information concerned with the written word from the visual cortex and converts it into sound-equivalent representations for decoding in Wernicke's area. [f.]

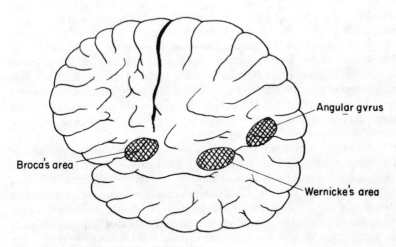

Language areas of the cerebral cortex

Latency period In Freudian theory, the period from the end of the *oedipal* stage around 6 years, until the onset of *puberty* and the beginnings of genital sexuality. Freud saw this as a relatively calm period of the child's development.

Latent content The underlying and usually hidden meanings in the account provided by a patient in *psychoanalysis*. The term is usually used about *dreams*, which Freud thought were particularly rich in indications of unconscious processes for anyone who could see past the *manifest content*.

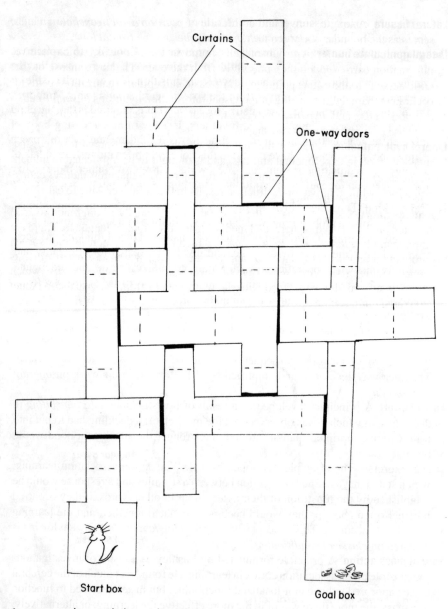

Curtains

One-way doors

Start box

Goal box

A maze used to study latent learning

Latent learning A system of *learning* first demonstrated in 1932 by Tolman, who presented clear empirical evidence that even laboratory rats could form internal, cognitive representations of a complex maze; and that learning need not necessarily show immediately in behaviour but might remain latent until it was advantageous to use it. Latent learning was important as a concept because it provided a counter to the *behaviourist* argument that learning and changes in behaviour were synonymous. [f.]

Lateral fissure A long fissure found at the side of each *cerebral hemisphere* which serves as the boundary between the *temporal lobe* and the *frontal lobe*.

Lateral geniculate nuclei A group of cells found in the *thalamus*, which receive information carried along the *optic nerve* from the eyes. The first synapse of the optic nerve is found at this point, and some basic perceptual organization seems to occur here, namely, the sorting of the visual information by means of *simple cells, complex cells,* and *hypercomplex cells*, such that hypercomplex cells fire in response to simple patterns and shapes.

Lateral hypothalamus A part of the thalamus which has been shown to affect the intake of food in experimental animals, and is thought to be implicated in human *eating disorders*. Electrical stimulation of the lateral hypothalamus induces eating behaviour, while removal or destruction results in the animal ceasing to eat.

Lateral thinking Thinking which involves a 'sideways leap' from conventional attempts to solve a problem, and which reaches a solution by adopting novel tactics or by reformulating the problem in an unusual manner. Lateral thinking has been promoted since the 1960s by de Bono, and involved a search for originality and flexibility in mental operations which would counteract sterile and hidebound problem-solving practices, both in management and in day to day *problem-solving*. *Divergent thinking* has a similar meaning. See also *creativity*.

Laterality Specialization of function on one side. Used both of *handedness* and of the specialization of function in either the left or right hemisphere of the brain.

Law of effect The principle developed by Thorndike, that a response which was followed by a pleasant consequence would be more likely to be repeated. This idea was developed and amplified by B.F Skinner, in his work on *operant conditioning*.

Law of effort A principle developed as a result of investigations into *imprinting* in ducklings, in which it was observed that the more effort a duckling had to put into following its imprinted 'parent' around, the stronger the attachment bond would become.

Law of exercise The principle developed by J.B. Watson, on association learning, which stated that a learned connection between a stimulus and a response would be established by the repetition of their association. In other words, if they occurred together often enough, they would become associated together, and the learning would have occurred. This concept was later developed more fully by Pavlov in his research on *classical conditioning*.

Law of mass action A principle formulated by Lashley as a result of investigations into the role of the association cortex in learning. He found that much of the cerebral cortex appeared to have non-localized functioning, but instead seemed to function as a mass: the more there was of it, the more effective the learning; or alternatively, the greater the amount destroyed, the greater the learning impediment. See also *equipotentiality*.

Law of Prägnanz The principle by which meaningfulness or organization of visual stimuli occurred, according to the *Gestalt* psychologists. The law of Prägnanz is concerned with the ways that perceptual organization occurred through the subsidiary principles of proximity, similarity, closure and good gestalt, such that meaningful figures against backgrounds are seen, rather than just a jumbled mass of disparate elements of visual information. [f.]

Lay epistemology

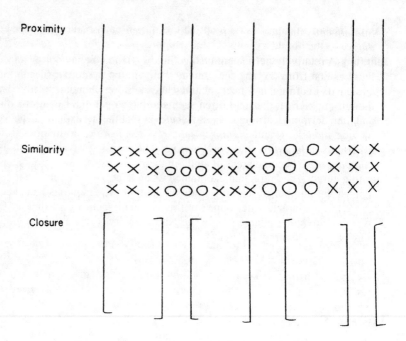

Proximity

Similarity

Closure

Example of the Gestalt Law of Prägnanz

Lay epistemology An approach articulated by Kruglanski (1980), lay epistemology represents an attempt to structure the processes by which *social attributions* and *social representations* become incorporated into the individual's personal knowledge frameworks and used as 'common sense'. One of the distinctive processes identified in studies of lay epistemology is that of 'freezing', in which the person latches onto one specific explanation and then does not change it even in the face of directly contradictory information.

Leadership style Patterns of behaviour by designated group leaders which have been found in empirical studies. One division is between task-oriented leaders whose efforts are directed to getting the job done, and maintenance-oriented leaders who pay more attention to ensuring that the group is working well together. Other forms of leadership are: authoritarian, authoritative (who maintain authority through example and negotiation), democratic (who work through persuasion and consensus), and laissez-faire (who largely leave the group to find its own solutions). Similar styles have been identified in studies of parenting. See *child-rearing styles*.

Learned helplessness A concept demonstrated experimentally by Martin Seligman. He showed that animals which had received unpleasant experiences about which they could do nothing, were less ready to undertake action when in a similar stituation where a relatively simple response would avert an unpleasant experience. Instead, the animals would remain passive, and do little to help themselves, not even struggling. Seligman drew parallels between the behaviours shown by animals in this condition and the behaviours associated with *depression* in humans. From these parallels he developed *helplessness theory*, which proposes that (some) depression may result from a belief of having no control over bad events.

Subsequently the theory was revised by Seligman and others in terms of *attribution theory*.

Learning A relatively permanent change in knowledge, behaviour or understanding that results from experience. Innate behaviours, maturation and fatigue are excluded. Learning has been claimed as the core phenomenon of psychology though in practice, the field often seems to have operated by producing a theory and then defining learning as being whatever that theory explains. Specialist areas include *modelling* and *imitation, motor skills, insight*, formation of *schemata, creativity, habituation* and *conditioning*. The learning of specific skills such as *language* have become areas of study in their own right.

Learning curve The graph that is obtained when a measure of competence is plotted against the number of learning trials the animal or person has had. The learning curve has a characteristic shape but this is usually achieved rather artificially by averaging together a large number of learning curves while the individual curves may be much less regular. [f.]

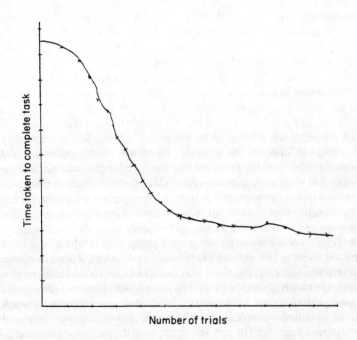

A learning curve

Learning set A generalized style of learning, or state of preparedness to solve problems in certain ways, which has been acquired through experience with similar types of problems. Possession of a learning set means that the individual is likely to look for that kind of solution in preference to any alternative strategy. Where problems are similar, learning sets may be advantageous, but may prove a hindrance to the individual faced with a problem which requires a novel approach.

Learning theory A theory about how learning occurs. Note that, as discussed under *learning*, the theory is not specifically about what learning is, for that tends to be assumed at the outset, though it may be modified as theory develops. Some theories, such as operant conditioning, are presented as accounting for practically all learning. Others deal with a particular type, for example *insight* learning. There are also theories for specific phenomena such as *transfer of training* and *modelling*.

Left hemisphere The left half of the *cerebrum*. The cerebrum is divided into two hemispheres by a deep fissure. In most people, the left hemisphere contains the *language areas*, and is also thought to be concerned with the general functions of logic and numeracy. It is generally referred to as the dominant hemisphere, as functions from the left hemisphere will usually override those from the right hemisphere.

Lesbian A female person who is sexually attracted to members of her own sex.

Lesion A term used to refer to damage to organic tissue which is usually used by psychologists to refer to brain or neural injury. Lesions may be surgical or accidental and may take the form of cutting of specific fibres or pathways, or of general damage caused, say, by the impact of a heavy object.

Levels of measurement Types of measurement which differ in how far they can be manipulated mathematically. The lowest level of measurement is known as *nominal data*, which is information which cannot be ranged on a scale, but only organized into different categories. The next level is of *ordinal data*, which are data which can be put into a definable order, and so can be ranked. However, no information can be provided about the size of the difference between any two items. For example, if colours are arranged in order of preference, it is possible to say that one colour is liked more than another, but it is not possible to be precise about how much more. The third level of measurement is *equal-interval data*, in which the measurements can be ordered on a scale which has equal intervals: e.g. measurements of temperature in fahrenheit or centigrade. The highest level of measurement is known as ratio data, which is equal-interval data with an absolute zero, such that it is possible to describe one score as a precise proportion of another. Because temperature in centigrade is only an interval scale there is no sense in which 40 degrees is twice as hot as 20 degrees. However height is a ratio scale, so 2 metres is twice as high as 1 metre.

Levels of processing A theory of memory put forward by Craik and Lockhart, which argues that information may be processed at a number of levels, depending on how it is organized, linked with other memories, tied in with emotional experience, and so on. Information which has been only superficially processed or accepted passively will be readily forgotten, and this is used to explain the phenomena of rapid forgetting previously characterized as *short-term memory*. Information which has been processed more deeply will be retained for a longer period of time. [f.]

LH See *lateral hypothalamus*. Also used as an abbreviation for 'left-handed'.

Libido A term originally used by Freud to refer to sexual energy which is derived from the *id* and is available to power mental and physical activity. Later, Freud regarded libido as a general life energy. In common usage, the connotation of sexual energy is still associated with the term.

Lie-detector See *polygraph*.

Life-event An event that results in a major change in the life situation of a person. There is evidence that all life events, even those that are fundamentally positive,

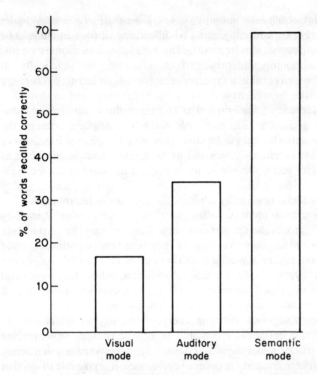

Levels of processing in memory

impose a *stress*. Holmes and Rahe have produced a 'Life Events Scale' which gives weightings to different events, ranging from 100 for death of a spouse down to 12 for Christmas and 11 for minor violations of the law. The scale can be used to provide a total score for all of the life events experienced during, say, the last year. People experiencing a lot of change will obtain a high score, and high scores may indicate that a person is at higher risk of illness or accidents. Negative life events may also make some people more prone to *depression*.

Life span The entire period of a person's life. There has been a move within *developmental psychology* to study the whole life span rather than restrict the field to childhood. This approach has opened up the adult years from 20 to 65 as an important period within which to study development, but so far the obvious worthiness of the objectives has not been matched by exciting findings or theories.

Light adaptation The process by which the *photosensitive* cells of the retina adjust to changing levels of illumination. By varying their sensitivity to light, and also by the adjusting of the pupil of the eye in order to maximize light intake in dim conditions and minimize it when conditions are bright, the individual adjusts their perception to accord with the amount of light available. See also *dark adaptation*.

Lightness constancy The experience of a consistent level of illumination in different environments even though objective measurements of the light available would show them to be widely varied. For example: sitting under electric light in the evening is often perceived as 'full light', and as equivalent to daylight, despite the fact that in reality the light level is several thousand times dimmer than sunlight. See also *perceptual constancy*.

Limbic system A general term used to refer to a series of small structures buried deep in the centre of the brain, which seem to be involved with several disparate functions, including the encoding of memories (the *hippocampus*); motivation (the septum); and emotion (the amygdala).

Linguistic To do with *language*. The term linguistics is used to refer to the study of language itself.

Linguistic relativity hypothesis Sometimes also known as the *Sapir–Whorf hypothesis*, this is the idea that thinking is dependent on the language used by the individual. In other words, that the possession of words for each *concept* shapes a person's thought. In the 'strong' form of the hypothesis, words are seen to determine thought entirely; but a 'weak' form has become more generally accepted, which states that the words available serve to facilitate and amplify thought, and to indicate relationships between concepts, rather than actually to determine them.

Lithium The basis for drugs which are an effective treatment for *bipolar depression* in most cases. The effect seems to be to prevent the manic phase, so that the cycle does not continue and so the depressive phase is also prevented. The method of action of the drug is not known. Lithium is an element that is close to sodium and potassium so it is thought likely that it changes neural transmission in the central nervous system. The drug has to be taken continuously and it is dangerous, with significant side-effects even in carefully controlled doses.

Localized functions Functions, usually of the cerebral cortex, which have been shown to be located at a particular site. Among the localized cortical functions are motor control, located in an area alongside the *central fissure*; body skin sensation, located on the other side of the central fissure; vision, located in the *striate area* or *visual cortex* in the occipital lobe; olfaction (the sense of smell), located in a strip at the base of the *temporal lobe*; and the *language areas*.

Location constancy The way in which the perceptual system automatically modifies its judgements and estimations of objects and distance, depending on their location and the location of the perceiver. Objects viewed from an alternative location are not perceived as having changed their position, despite the fact that the background to them has altered: rather, they are seen as having remained constant. The *perceptual constancies* are often used to illustrate the way that the received visual image is only a part of perception: what is known on a cognitive or experiential level is an equally important part.

Locus of control A concept at the core of a social learning theory developed by Rotter in the 1950s. It refers to the belief that a person has about where social reinforcements originate: whether they are internal to the person, or external. Someone with an internal locus of control (LOC) will tend to believe that marks on an essay depend on the amount of effort and ability applied to writing it. Someone with an external LOC will tend to attribute the marks to luck, predestination, or the whims of the person doing the marking. LOC can be measured using a variety of short self-report scales and has been found to relate meaningfully to how people behave in a great variety of situations. Such evidence supports the construct *validity* of the scales. Writings in the area often imply that an internal LOC is preferable. It is true that an internal LOC is more likely to result in the individual making efforts to improve their situation but whether this is useful depends on whether events are actually under their control or not. A similar but not identical concept was developed more or less independently in *attribution theory*. See *internal–external scale*.

Logic A set of rules by which conclusions can be reliably deduced from initial statements (propositions). Logic can be applied without regard for the truth of propositions. For example, 'All students work hard and those who work hard pass their exams, therefore all students pass their exams' sounds logical. The fact that it is not true that all those who work hard pass their exams, means that the conclusion is not necessarily true, although it could be, by accident. Logic has been of interest in psychology because it can be regarded as perfect reasoning and is therefore a starting point for analysing how people reason. It turns out that people are much more sophisticated and rather less rigid in their thinking than any logic that has been invented, and there is not too much similarity between the two processes.

Longitudinal study A study which takes place over a period of time, and is concerned with studying some form of development or change which occurs over time. Longitudinal studies have been valuable in challenging many erroneous or commonly-held beliefs. For example, longitudinal studies of the relationship between ageing and *intelligence* suggest that intelligence, if used, continues to develop and increase throughout life rather than declining with age as was once thought. See *cross-sectional study*.

Long-term memory (LTM) A term used to describe memories other than those which remain for a few seconds only. According to the *two-process theory* of memory, any information which is retained for any length of time above a few seconds is deemed to have been stored in LTM, while that which lasts just for a brief interval (such as a telephone number which has just been looked up) is considered to have been stored in *short-term memory*. Many modern researchers question this commonly accepted distinction, arguing that it is unnecessary and that it fails to discriminate between information retained for varying periods of time. One alternative to this approach has been the *levels of processing* theory, which argues that the decisive factor in how long information is retained is how deeply it has been organized and processed and that there is no need to postulate separate memory stores.

Love need A term used by some humanistic psychologists to refer to the need for affection or *positive regard* from others, which is seen as a fundamental part of human nature.

LSD See *lysergic acid diethylamide*.

LTM See *long-term memory*.

Lysergic acid diethylamide A recreational drug which forms a potent *hallucinogen* when ingested, producing visual disturbances, sometimes hallucination, and a heightened or distorted awareness of reality.

M

MA See *mental age*.

Magical thinking The belief, common in young children, that thinking of something makes it happen or exist.

Main effect The overall relationship between a class of independent variable and the dependent variable. The term is used mainly in *analysis of variance*.

Major hemisphere See *dominant hemisphere*.

Maladjustment A poor *adjustment*. The term is used of people, particularly children and adolescents, whose behaviour is judged to conflict strongly with the expectations and requirements of society.

Mania An emotional disorder during which there is elation, talkativeness, impatience with others, over-confidence and an uncontrolled flight of ideas. See *bipolar depression*.

Manic depression An emotional disorder in which there is an alternation between mania and depression. See *bipolar depression*.

Manifest content The overt content of an account, usually of a dream. Dream interpretation involves seeing beyond the manifest content to understand the underlying meaning: the *latent content*.

Manipulative skill A skill which involves direct action with the hands, usually in terms of handling and placing of objects.

Mantra A word or phrase on which a person concentrates as an aid to *meditation*. Traditionally the mantra is derived from Hindu scripture and has spiritual power.

Marijuana A *psychoactive drug* which induces a feeling of lethargy and relaxation when consumed. Marijuana is derived from the cannabis plant, and may be consumed either by smoking the dried leaves or resin of the plant, or by eating small pieces of the resin. The use of marijuana as a relaxant is extremely common in many areas of the world, including Africa, the Middle East, and Central America. It is widely (although often illegally) used as a recreational drug in Western industrial societies. Marijuana appears to have its main effects by increasing the *noradrenaline* levels in the brain.

Masochism Obtaining sexual gratification from pain or humiliation. Often associated with *sadism*.

Massed practice Extended periods of practice while learning a new skill, taken without breaks. Massed practice has been found to be less effective than *distributed practice* which allows for consolidation.

Mastery play Play during early childhood which leads to the acquisition of new skills. This definition leaves open the question of whether children are motivated to achieve mastery, or perhaps cannot avoid learning when having fun. See *play*.

Matching The name given to ensuring that two sets of experimental materials or subjects are identical in all important respects. A matched task or test has questions carefully selected to ensure that, in each test, the questions are equivalent in difficulty, and in the type of problem posed. It is usual to select a group of subjects matched in terms of age, sex and overall intelligence levels, although other criteria may be used if required for the study.

Maternal deprivation A concept proposed by John Bowlby and Rene Spitz to account for the poor development of children brought up in institutions. Of the various disadvantages suffered by these children, the theories of the time, the 1940s, focused on the lack of consistent mothering. Bowlby added other evidence and concluded that any disruption of mothering, especially between 6 months and 3 years, was likely to have damaging long-term consequences. The belief that infants should never be separated from their mothers became stressed beyond anything Bowlby had claimed. It has been suggested that the concept of maternal deprivation was exploited in order to remove women from employment and so release jobs for men at the end of World War II. If so, then similar calls might be expected during any other period of high male unemployment,

The concept of maternal deprivation was soon challenged, and much evidence has now been accumulated showing that good development is possible without the consistent presence of a mother or mother substitute. However, the evidence does not show that good development is especially easy under these circumstances. A fair statement might be that, while around 1950 mothering could be thought of as a thing that the infant either got or did not get, we now know that the normal processes of mothering provide a great variety of physiological and emotional effects, learning experiences, motivations, practice in social interaction and no doubt much else besides. Substituting for all of these may certainly be possible but it is likely to be difficult. See *attachment; maternal privation.*

Maternal drive The tendency, usually presumed to be innate, to engage in caretaking behaviours such as nest building, retrieving and suckling during the infancy of offspring. The tendency is displayed by mothers, and sometimes by fathers, in many species. Use of the term *'drive'* implies that there is some basic need to be maternal, an assumption that should not be accepted uncritically. The term 'maternal instinct' is sometimes used instead, but this is even more likely to bring in assumptions for which there is inadequate evidence. The most misleading use of the terms comes when meanings which have been built up by studying species like rats are applied uncritically to humans.

Maternal privation Rearing from birth without a mother. Strictly, privation means never having, while deprivation means having something taken away. Experiments of total maternal privation have been carried out on various species, though not with humans. However these are typically classed as maternal deprivation studies, and in practice the term *maternal deprivation* is used for all variations of a shortage of mothering in the upbringing of young.

Maturation The term used to describe behavioural or physical changes which occur as a direct result of *genetic* action, but which emerge as the animal or human matures, or grows older. A clear example of maturation in terms of physical development are the changes which occur at puberty. In the 1920s Gesell proposed a theory that nearly all development is controlled by maturation and so is independent of practice or experience. It is still thought by some psychologists that the development of much behaviour may be maturational.

MBD See *minimal brain dysfunction.*

Mean The name given to the arithmetic average of a set of numbers, calculated by summing the numbers and dividing this total by the number of figures in the set. The mean is one of the three main *measures of central tendency*, but it can only be used for equal interval or ratio *levels of measurement.*

Measures of central tendency A collective term for all of the statistical measures which tell you something about the middle of a distribution of scores. The mean is

Measures of dispersion

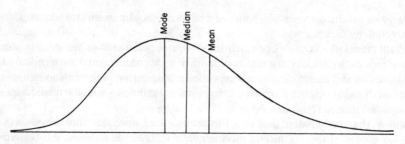

Measures of central tendency in a skewed distribution curve

the most widely used but others may be more informative in certain circumstances. For example: when considering the spread of incomes in a particular country, the *mean* may be unduly affected by a few extremely rich people. The *median* will tell you the income of someone right in the middle of the earning population, while the *mode* will tell you the most common income. [f.]

Measures of dispersion A collective term for all of the statistical measures which tell you something about the way a distribution of scores is spread out. See *range, standard deviation, variance*. [f.]

Median A *measure of central tendency* which is calculated as the middlemost score from a given set. 50% of the scores in a given set will fall at or below the median score, and 50% fall at or above it. The median is appropriate for use with *ordinal levels of measurement*.

Mediators Processes (e.g. *memory, perception, thought*) that come between a stimulus and a response. The early *behaviourists* claimed that as mediators cannot be observed directly, they should not form part of scientific psychology. *Cognitive*

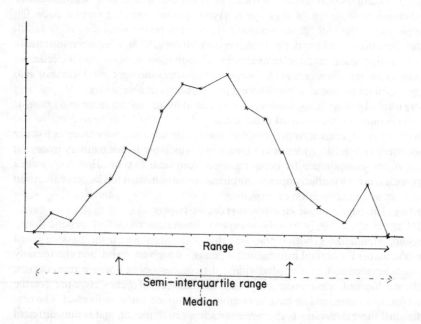

A frequency polygon showing median and semi-interquartile range

psychologists on the other hand regard them as the main subject matter of psychology. See *schema*.

Medical model An overall approach to abnormal behaviour or personality which assumes an individual organic source of any disorder, mental or physical. The medical model has particularly been called into question in the explanation of the less serious *psychiatric disorders*. One of several problems with the medical model is that it tends to result in *labelling*.

Medulla The lowest part of the brain, formed by a thickening out from the spinal cord. Also known as the brainstem, the medulla mediates the *autonomic* functions of breathing, digestion, heartbeat and blood pressure.

Meiosis The process of cell division involved in *sexual reproduction*, in the formation of gametes (ova and spermatozoa). Unlike the kind of cell division involved in growth and tissue repair (see *mitosis*), this involves the separation of pairs of *chromosomes*, such that the resulting cells are haploid (have only half the normal number of chromosomes). In order to form a complete *zygote*, which can develop to form a new individual, these cells must combine with another haploid cell to make up the full complement of chromosomes. In this way, the newly formed individual comes to inherit half of its chromosomes from each parent.

Memory The general term given to the storage and subsequent retrieval of information. Memory has been intensively studied by psychologists throughout the history of psychology, and consequently involves an extensive range of theoretical approaches and fields of enquiry. These include the study of episodic memory, everyday memory, *levels of processing, encoding* and *representation*, and physiological correlates of memory. See also *two-process theory of memory*.

Memory span A well-known measure of an individual's capacity for retaining small units of meaningless information over a brief period of time. In a typical measure of memory span, a list of digits is read out to someone at a regular pace. On completion of the list, the individual is required to repeat what they have heard, either forward or backward. First observed by Miller (1955), it has been repeatedly observed that the average span available to the individual is of 7 digits, ± 2; and that this can only be increased by some system for *chunking* the information into meaningful units. See also *two-process theory, levels of processing*.

Memory trace In older texts sometimes referred to as an engram, a memory trace is a hypothetical 'image' of what is to be remembered, which has been encoded and which is stored, for varying periods of time. The term memory trace is usually associated with the decay theory of forgetting, which holds that memory traces die away if not strengthened by being recalled from time to time. However, as this approach is not particularly open to empirical investigation, it has largely fallen into disfavour as an explanation of forgetting.

Menarche The beginning of *menstruation* during puberty.

Mendelian genetics The currently accepted theoretical model of genetic transmission, Mendelian genetics proposes that it occurs through the passing on of discrete units of inherited information – genes – which are fixed, and change only through accidental mutation. Individual differences occur because reproductive cells are haploid, containing only half the number of genes required for the complete organism, and so have to combine to produce a new individual. The new individual therefore inherits characteristics from each parent, and is thus different from either. The combination of Mendelian genetics with Darwinian evolutionary theory proposes that those combinations and accidental mutations which are favourable to the individual, in terms of helping it to survive, will be passed on because that individual will then become fitter, healthier etc. and therefore more likely to reproduce successfully. See also *Lamarkian genetics, genetics, evolution, sexual reproduction.*

Menstruation The phase of the monthly menstrual cycle in which, if the woman is not pregnant, the blood and other material which has built up in the uterus following ovulation, is discharged. Many cultures have beliefs about the dangerousness of women during or around menstruation. The major Western version concerns premenstrual tension.

Mental age A construction developed by Binet in his early work on the measurement of intelligence, mental age refers to the abilities of the individual by comparison with others of that society. By selecting a series of age-appropriate problems and tasks, a set of age *norms* is developed, allowing each child to be assessed in terms of how far they measure up to these criteria. The level of difficulty of items at which the child starts to fail is compared to the norms. The average age of children who pass the items up to this point is found, and this is regarded as the mental age of the child being tested. Binet's original formulation of the *intelligence quotient* involved the comparison of mental age with the child's *chronological age* ('real' age).

Mental handicap A general term for people of limited intelligence as measured by intelligence tests. The classification of people in these terms raises many difficulties as it labels them in terms of a particular aspect of human ability. Worse, it is an aspect which is highly valued in this culture, and it is measured by tests which many regard as unsatisfactory. In particular, *intelligence tests* are usually standardized on the 'normal' population and they may have less *validity* when applied to other groups. However, people who are mentally handicapped need to be identified so that they can be provided with special resources at an appropriate level and at present, intelligence testing is the most reliable and accurate way of achieving this.

The term 'mental handicap' is one in a long line of *labels* which may have been scientifically neutral when first used but which tend to become unacceptable as they pass into the general language as terms of abuse. For example, an earlier grading of mentally handicapped people was as idiots, imbeciles, and morons. The term 'mental deficiency' is sometimes used, particularly for mental handicap which is believed to be due to brain damage. The term's current unpopularity is in part due to a recognition that it is not very productive to attempt to distinguish between organic and non-organic cases. 'Mental retardation' is also widely used, particularly in the American literature. More recently, 'special learning difficulties' has replaced 'mental handicap' as the official term, although this solution also raises difficulties, notably in undervaluing the efforts demanded of those who care for people of very limited intelligence.

Mental imagery The use of imagined pictures, or other sensory images, such as sounds or smells, to represent information in the mind. Mental imagery involves recreating the apparent sensation, as part of the process of memory or thinking. See also *symbolic representation.*

Mental retardation A general term for limited intelligence. The term tends to carry a misleading assumption that the low intelligence is due to either a slowness of mental functioning, or slow intellectual development. See *mental handicap* for a fuller discussion.

Mental set A state of preparedness to perform certain kinds of mental operations rather than others. Mental sets may refer to particular kinds of problem-solving (see *learning set*), or to readiness to perceive certain things rather than others (see *perceptual set*), or to a preparedness to remember certain items of information in preference to others.

Meta-analysis A research technique which involves comparing the outcomes of a number of different studies in the same area, and examining the general themes or principles which can be identified as a result.

Metacognition An overall term used to refer to the knowledge about how *cognitive* processes work which is often highly influential in cognitive development. The study of metacognition includes the study of the ways in which people monitor and control their own cognitive activity, such as being aware of cognitive limitations (knowing that you don't know), or abilities (knowing that you can learn certain types of information readily). The act of looking a word up in a dictionary, for instance, is one which would be unlikely to happen without metacognition.

Metalinguistic awareness Knowledge about the nature, forms and functions of language. It is possible to be a fully competent language user without metalinguistic awareness, but the different ways that people understand how language works are likely to influence how they interact with their world and each other. It is therefore an important area of study for psychologists.

Metamemory Knowledge about how one's *memory* works, or what its limitations are. Such knowledge often directly affects behaviour, such as a decision to write a note to yourself to remind you of something, or to adopt a specific revision technique to make remembering easier. See also *metacognition*.

Metapelet The name given to a child-nurse or professional carer for children in an Israeli *kibbutz*. Such an individual carries the responsibility for the care of the children, rather than the parents and oversees their day to day experience and early learning.

Method of loci A *mnemonic* technique in which a mental image is formed which visualizes items to be remembered at specific locations. Usually the locations take the form of landmarks along a familiar walk, or journey – something which is already well known to the person forming the image. By subsequently visualizing the journey, the individual is reminded of the items to be remembered.

Micro-electrode recording A means of investigating neural activity, by recording the firing of single *neurones*. It consists of a technique whereby microscopic electrodes which are sensitive to very small electrical charges are inserted into the appropriate region of the brain or nervous system. These electrodes record when their target cells fire. By means of this technique several discoveries have been made, including the processing of visual information in the *thalamus* and the *visual cortex* and the neurones involved during the *imprinting* process in young chicks.

Midbrain A part of the brain above the *brainstem*, which includes part of the *reticular formation* and the *pons*, and seems to be active in the integration of sensory input and motor activity.

Middle ear The air-filled chamber of the ear which is separated from the outer canal by the tympanic membrane, and which serves to amplify the received signal, in preparation for its *transduction* in the *inner ear*. The middle ear contains three small bones, known as the ossicles, which form a link from the tympanic membrane at one side of the chamber to the oval window at the other. Each ossicle receives the vibrations in turn, and amplifies them slightly as it passes them on. In sequence, the ossicles are the malleus (hammer bone), incus (anvil bone) and stapes (stirrup bone), named in accordance with their overall shapes.

Minimal brain dysfunction (MBD) The preferred choice from a number of terms which have been proposed to account for, or at least label, a set of quite common childhood conditions. The conditions include *hyperactivity, attention deficits* and *clumsiness*. They are the kind of problems that could arise because of brain damage, but no organic damage can be identified in these children. The conditions were therefore classified as 'minimal brain damage' with an implication that there was damage but it was too minimal to be detected. As psychologists came to realize that invisible brain damage was not really a useful explanation of anything, alternative terms were proposed such as minimal cerebral injury and eventually MBD. It is now becoming recognized that the various conditions have little in common, so the search for a suitable term under which they can all be grouped is likely to be abandoned.

Minimal group paradigm An approach to the study of *social identification*, in which minimal indicators of group membership are shown to produce reliable social effects. In a typical example, participants in a minimal group paradigm experiment are allocated to membership of a group according to some arbitrary criterion, such as the toss of a coin. When asked to allocate resources to members of their own or other groups, they then show a reliable tendency to favour their own group above the others. Minimal group studies have generated a number of hypotheses about social identification which have been supported by more realistic investigations, including the tendency to accentuate differences between the in-group and the out-group, and to stereotype out-group members. However, such studies appear to be particularly susceptible to *demand characteristics*, as it is difficult to imagine what other behaviour could be expected of the co-operative subject whose only information is that they belong either to one group or another.

Minnesota multiphasic personality inventory (MMPI) The most famous question-naire measure of *personality*, consisting of 550 items and providing eight scales or traits. The objective evidence suggests limitations in its use either as a clinical predictor or as a guide to how people are likely to behave in practice. It is, however, still widely used in research.

Minor hemisphere The name given to the half of the cerebrum which does not form the *dominant hemisphere*. In most cases, this is the right hemisphere, but in some people the right hemisphere is dominant, and the left one forms the minor, or non-dominant, hemisphere.

Mitosis The process of cell division which results in each new cell possessing a full complement of chromosomes: an identical copy of the genes carried by the parent cell. This is the most common form of cell division, being the type which is involved in tissue growth and repair; it contrasts with the form of cell division involved in *sexual reproduction* which is called *meiosis*. Mitosis also made possible the development of *cloning*, in that since each cell of the body carries the full genetic complement of that animal, given the right medium for cell division and growth, it is possible to recreate an identical animal from a cluster of parent cells.

Mnemonic An aid to memory, which can be achieved in any way, including leaving a note for oneself. There have been several different kinds of mnemonics identified and developed over time. Many of them have to do with forming of mental images which will help the person to remember connections between items, or lists. Some mnemonics rely on the use of visual imagery, such as the *method of loci* or the *key word technique*. Other mnemonics rely on verbal processing, such as first-letter mnemonics, in which the first letter of each item spells out a new word or a sentence. For example 'Richard Of York Gave Battle In Vain' for the colours of the spectrum (red, orange, yellow, etc.). The famous 'knot in the handkerchief' is a mnemonic which combines visual and tactile cues to help the person to remember.

Mode The most frequently occurring score within a distribution. One of the *measures of central tendency*.

Modelling Providing an example which can be imitated, such that the imitator is able to learn new styles of behaviour. Modelling is considered to be an important aspect of *social learning* in children, because what is copied is more general than the imitation of a specific behaviour. It is often used explicitly in therapy, to allow adults to vary their styles of interaction with others.

Modes of representation Ways of coding information internally. Bruner identified a developmental sequence in representation, arguing that the first mode to develop was *enactive representation*, in which information is stored as 'muscle memories'. As the child's experience widens, and the environment makes increasingly complex demands, more sophisticated modes of representation are required: first *iconic representation* (using images) and then *symbolic representation* (in which information is represented by symbols).

Mongolism See *Down's syndrome*.

Monochromatism Seeing in one colour only; usually interpreted as seeing in black and white. In other words, monochromatic individuals are those who are entirely blind to all wavelengths of colour, a rare condition, as most *colour blindness* involves lack of sensitivity to a few wavelengths only.

Monocular depth cue An indication of how distant something is, which can be detected just as well with only one eye as it can with two. Monocular depth cues include relative size, height in plane, superposition, *gradient of colour, gradient of texture*, shadow, and motion parallax. [f.]

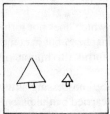
Relative size

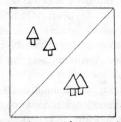
Height in plane/superposition

Shadow

Some monocular depth cues

Monotropy Bowlby's original idea of the way in which *attachment* develops between the young infant and its mother. Based on ideas from *ethological* studies of *imprinting*, the theory stated that the relationship which an infant formed with its mother was qualitatively different from any other relationship which it formed with other people, and that if the bond was broken, through separation, during the early years of life, then the child could suffer permanent damage. This led to the *maternal deprivation* debate, and produced extensive research into attachment and *mother–infant interaction*.

Monozygotic twins *Identical twins* who have developed from the same fertilized *ovum* which has subsequently split to develop as two independent foetuses. Monozygotic twins are identical genetically, and hence have been used in studies of the relative importance of genetics and environmental influences in development. But see *heritability*.

Moral anxiety In Freudian theory, anxiety that comes from a fear of the *superego*. As the superego has incorporated the rewards and punishments of the parents, it is able to inflict pain, and if it becomes too powerful the person may live in a chronic state of anxiety. See also *neurotic anxiety*.

Moral development This should refer to the development of moral standards and behaviour. In fact the term has been taken over by a particular approach which concentrates on moral judgement. Piaget analysed tendencies in the developing moral judgement of the child, such as a progression away from a belief in immanent justice. Lawrence Kohlberg developed Piaget's ideas and produced a scheme of six stages of moral reasoning along which the child progresses. While moral reasoning is important, the theory has been criticized both for the ways in which the stages are defined and for appearing to undervalue other aspects of moral development such as moral behaviour. See *autonomous morality, conventional morality, pre-moral stage*.

Moral realism Another name given to the stage of *heteronomous morality* described by Piaget, in which the child accepts fully the rules which are given to it by society and those in authority. See also *autonomous morality*.

Moro reflex A reflex found in newborn babies in which the limbs are closed in to the body and the hands are clenched. The probable function of the reflex can be seen when it is elicited by letting the baby slip while holding it and the *moro* reflex causes the baby to cling on to its caretaker. (Do not try this, take our word for it.) See also *reflex*.

Morpheme A unit of spoken language, in which basic speech sounds (*phonemes*), have been combined to produce basic syllables or simple words. A morpheme is the smallest unit of speech to have any real meaning in communication.

Morphology The study of form, or complete units. In linguistics, morphology refers to the study of how *morphemes* are utilized and combined in speech; in biology it refers to the study of the form and function of parts of the anatomy, or the structure of the living being.

Mother–infant interaction The forms of interaction between caregivers (who may or may not be the mother) and infants, particularly in the first few months of life. This interaction has been extensively studied to provide information about the beginnings of *attachment* and has been found to be very complex. Often called parent–infant interaction for obvious reasons.

Motherese A simplified form of speech that adults adopt when talking to very young children.

Mothering Providing the physical, cognitive and emotional care and stimulation required by an infant or child. Research indicates that this kind of care can be provided by any adult or older child provided he/she has an appropriate commitment, a knowledge of the needs of infants, and an ability to respond to the signals offered by the infant. There is therefore no sound reason to suppose that this care can only be provided by the biological mother or by a woman.

Motivation The general term given to an inferred underlying state which energizes behaviour, causing it to take place. There has been extensive physiological research into the neural mechanisms involved in motivational states such as hunger, thirst, the need for sex, exploration of novelty and so on. In addition, much research has emphasized the social aspects of motivation: the *need for positive regard* from others; or the way that specific forms of behaviour may occur as a result of the need to communicate or interact in meaningful ways with other people. While the majority of psychology textbooks limit discussions of motivation to physiological factors and need theories, a more comprehensive formulation of human motivation might incorporate a wider range of motives. These would include motivations arising from *cognitive* processes, such as *cognitive dissonance* or *personal constructs*; factors involved in motivating personal action such as *self-efficacy beliefs, locus of control, attributions* and *learned helplessness*; affiliative motivators such as *empathy*, or *positive regard*; and socio-cultural motivators such as *social identifications* and *social representations*.

Motive A specific inferred reason put forward to explain the likelihood of a particular behaviour occurring. See *motivation*.

Motivated forgetting A term for the forgetting of information as a result of an *unconscious* unwillingness to remember it (e.g. the forgetting of an impending dental appointment, because you don't want to go). According to Freud, all forgetting was motivated forgetting in some way: either because it could lead to the recall of deeply buried childhood traumas, or because the information which was forgotten was symbolic of such trauma. Other researchers identified alternative explanations of many kinds of forgetting, but motivated forgetting is still considered valid as an explanation of some instances of failure to recall information.

Motivators Specific incentives or aspects of the environment which can induce certain forms of behaviour in the individual. The term has been commonly used in *management theory*, where it includes such items as the provision of personal career development for individuals at work, or bonus payments which would encourage those in employment to work harder.

Motor area The part of the cerebral cortex which is directly concerned with the mediation of physical actions. This area forms a strip running alongside the *central fissure*, on the side of the frontal lobe. It runs directly parallel with the *somatosensory area*, and, in a manner similar to that of the organization of the somatosensory area, different parts of the strip mediate activity in different parts of the body. The most mobile parts of the body, such as the hands, have a large proportion of surface area representing them in the motor area.

Motor end plate The part at the very end of a *motor neurone* where the *axon* divides into small *dendrites*, which spread out and make *synaptic* connections with *receptor sites* in the muscle fibres. The *neurotransmitter* involved at the motor end plate is *acetylcholine*.

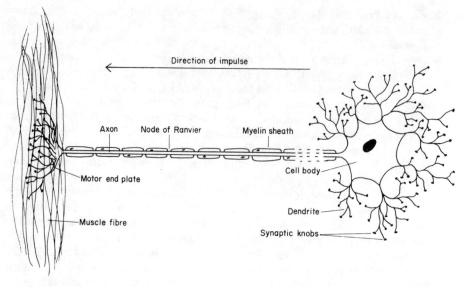

A motor neurone

Motor neurone A nerve cell which transmits information in the form of electrical impulses from the *central nervous system* to the muscles of the body. This information forms a signal for muscular contraction, resulting in movement of the limbs or body. Motor neurones tend to have their cell body located within the *grey matter* of the central nervous system itself, surrounded by dendrites which receive information from many other neurones. The *axon* is elongated, and reaches from the central nervous system to the muscle fibre itself, where it spreads into *dendrites* to form the *motor end plate*. Motor neurones are usually *myelinated*, which speeds up the passage of the impulse along the axon and allows more accurate timing. See also *sensory neurone, connector neurone*. [f.]

Multicultural Involving characteristics and aspects of several different cultures simultaneously. It is usually used to refer to modern societies, in which members of several cultural groups live, each bringing aspects of their previous culture to bear on the life of the society.

Multiple mothering Child care which is carried out by a number of different people, usually in succession. Infants in institutions were often exposed to a succession of caregivers and it is widely accepted that this form of *maternal deprivation* resulted in long-term difficulties in forming relationships. These days considerable efforts are made to avoid the repeated making and breaking of *attachments* in children who have to be brought up in care.

Multiple personality A rare condition in which a person functions with two or more distinct *personalities*. The personalities may alternate and may seem to be quite unaware of each others' existence. Multiple personality is not a form of *schizophrenia*, but a development of a phenomenon which is quite common and normal in childhood. See *dissociation*.

Multiple sclerosis (MS) A progressive degenerating illness which results in the person gradually losing motor co-ordination and control. MS is produced by the destruction or degeneration of the *myelin sheaths* covering the axons of nerve cells

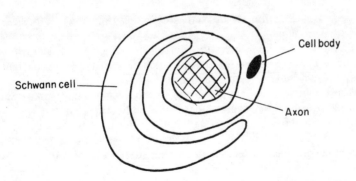

A myelin sheath in cross-section

in many parts of the brain, thus slowing down the transmission of information from one part of the brain to another. The process by which this happens is not yet fully understood.

Multipolar neurones See *connector neurones*.

Myelin sheath An insulating fatty substance which is wrapped round the *dendrons* and *axons* of neurones in the central nervous system. The myelin sheath is formed by *Schwann cells*, which coil themselves round the axon, thus preventing ionic transfer between the inside of the neurone and the surrounding fluids. A small gap between each Schwann cell is known as the *node of Ranvier*, and it is at these points that ionic transfer takes place. Because of this arrangement, the *electrical impulse* travels along the neurone in a series of jumps, which forms a much faster method of passing information the length of the *neurone* than would a gradual progression. This system is particularly utilized in the central nervous system itself, and in the receipt of sensory information and the transmission of motor impulses. In cases where a slightly slower progression is not a disadvantage, such as in the *autonomic nervous system*, neurones tend to be unmyelinated. The *white matter* of the central nervous system consists of the packed masses of myelinated nerve fibres. [f.]

Myelinated Covered with a *myelin sheath*.

Mz twins See *monozygotic twins*.

N

Nano- One billionth. One nanometer (nm) is a billionth of a meter and is used as a measurement of the wavelength of light. One nanosecond (ns) is a billionth of a second and is most likely to be encountered in measurements of the speed with which a computer can perform its simplest operation (called the cycle time).

Narcissism A love of the self. The more puritanical approaches to therapy regard narcissism as always undesirable, and when most of a person's affections are fixated upon themselves this must be so. However there is plenty of evidence that a healthy degree of affection for the self is essential to maintain *self-esteem* and productive functioning.

Narcolepsy A condition in which the person is subjected to sudden, short, uncontrollable episodes of deep sleep. It is much more extreme than the tendency to sleep during psychology lectures and also much rarer.

Narcotic Drugs which have both *sedative* (encouraging sleep) and analgesic (relieving pain) properties; usually of the opiate family, such as morphine or heroin.

Nativist An individual or school of thought holding that the important determinants of development are directly inherited, through genetic transmission. The name implies that the emphasis is on qualities which are inborn. Although nativists do recognize that environmental factors may have an effect on development, they consider such effects to be minimal, with the main explanation for individual differences being the *genotype* of the individual. The *maturational* theory of Gesell is an example of a nativist position. See also *empiricist*.

Natural selection See *evolution*.

Nature–nurture debate The name given to two opposed theoretical stances common within psychology. One stance emphasizes 'nature', the inheritance of abilities or characteristics, while the other emphasizes 'nurture', learning or the effect of environmental influences. Nature–nurture debates represent a convenient way of organizing some theoretical issues within psychology, but can often be deceptive in that they may present a false dichotomy, since almost every feature of humans has both a genetic and an environmental component. See also *nativist, empiricist*.

Naughty teddy A semi-legendary character introduced to modern psychology during a series of investigations of the effects of context in Piagetian *conservation* tasks. In investigations by McGarrigle, the changes in shape of the experimental substances were caused by a small teddy bear who 'lived' in a box on the experimenter's table and would periodically emerge to alter the experimental materials. The small children being tested had little trouble recognizing that the materials had, in fact, conserved their volume or number despite the actions of the toy bear. The studies were interpreted as throwing some doubt on the basic Piagetian assumptions concerning children's logical capacities. It was argued that the Piagetian findings resulted from the abstract nature of the conventional tasks and their lack of context, rather than from the child's inability to reason.

Necker cube A reversing figure, which appears to change its orientation irrespective of the intentions of the observer. The Necker Cube was cited by Gregory as an example of *hypothesis testing* in perception: without perceptual cues to indicate which way round the figure should be seen, the brain alternates from one plausible interpretation to the other. [f.]

Need A state of physiological deficit. Many needs, such as thirst, are associated with a *drive* or *motivation*, but others such as a need for vitamin C are not. The term has been extended to non-physiological needs such as *affiliation* and *achievement motivation*. See also Maslow's *hierarchy of needs*.

Need for achievement A proposed psychological need sometimes called N-Ach. See *achievement motivation*.

Negative aftereffects A set of illusions which occurs immediately after continuous or very intense stimulation of the visual system with the same sensory information. Possibly as a consequence of the *habituation* of *sensory neurones*, the opposite experience to the previous stimulation is experienced. The best-known negative aftereffects occur as a result of looking at something very bright, such as a

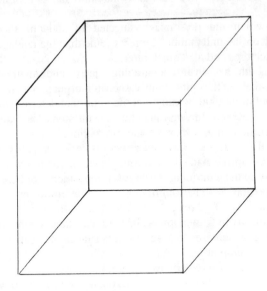

The Necker cube

lightbulb or the sun. For some minutes afterwards, the shape is perceived as a closed figure in the field of vision, and it is usually of the opposite colour to that which was seen. Colour aftereffects can also be induced by staring at a brightly coloured object for a couple of minutes, and then transferring the gaze to a plain background. Negative aftereffects also occur with movement: the *waterfall effect* is where a subjective impression of reversed movement is experienced after continuous exposure to movement in just one direction.

Negative incentive An object which has the opposite effect to an *incentive* so that the organism works to avoid or prevent it.

Negative reinforcement Reinforcement which involves the withdrawal or avoidance of something unpleasant, or aversive. Behaviour which has been strengthened by negative reinforcement – especially in the case of avoidance learning – is extremely resistant to *extinction*. The term is often wrongly applied to punishment. Negative reinforcement, like all reinforcements, strengthens the probability of a behaviour; *punishment* reduces or suppresses the target behaviour.

Neglect A standard category of *child abuse* indicating a substantial failure to provide what the child needs. Neglect is a form of passive abuse, and may involve poor physical care, a lack of cognitive stimulation or inadequate emotional warmth (see *attachment*). Neglected children are often also actively abused.

Neo-behaviourism A revised form of behaviourism in which it is recognized that *cognitive* processes play a role in determining behaviour.

Neo-Freudians A term used to describe psychoanalytic theorists who accept Freud's basic ideas, but have developed them further, often emphasizing social and cultural factors in psychodynamic processes. The British neo-Freudians have concentrated on *object relations* which in turn has made the study of *attachments* an important part of developmental psychology.

Neonate A newborn. For humans the neonatal period is usually taken to extend from birth to one month. Recently it has been recognized that the first major change in functioning occurs at around 8 weeks, and it has been suggested that the neonatal period should be extended up to the time that these changes start.

Neoteny The evolutionary model proposed by the biologist Stephen Jay Gould (among others) which argues that human infants are born 'prematurely', in the sense that they are far more helpless than most other young animals and thus have a long period of dependency before they are capable of independent existence. This extended dependency period allows biologically for more extensive brain development, and psychologically for an extended learning period, so providing the human being with both the ability to adapt to numerous different types of environment, and a highly developed capacity for learning.

Nerve A fibre or system of fibres which conveys sensory information from the sense receptors to the *central nervous system*, or motor impulses from the central nervous system to muscle fibres. *Afferent nerve* fibres consist of the *axons* or *dendrons* of *sensory neurones* bunched together to form a thread-like structure; while *efferent nerve* fibres consist of the axons of *motor neurones* arranged in a similar manner.

Nerve cell See *neurone*.

Nervous breakdown A non-technical term for a more-or-less complete loss of ability to cope with day to day living, showing itself in changes from the person's normal behaviour, such as extreme weepiness or anxiety, and general loss of psychological well-being.

Nervous system The network of nerve fibres which runs throughout the body, and includes the two main structures of the *central nervous system*: the *brain* and *spinal cord*.

Neuroanatomy The study of the structure and composition of the nervous system.

Neurochemistry The study of the chemical aspects of the nervous system, which includes the study of specific *neurotransmitters* and of ionic transfer within the *neurone*.

Neurone A cell which receives or relays information within the nervous system. The information takes the form of *electrical impulses*, which are passed from one cell to another by means of *synaptic transmission*. There are generally considered to be three main kinds of neurones: *sensory neurones*, which receive information from the sense receptors and pass it to the central nervous system; *motor neurones*, which transmit information from the central nervous system to the muscles, thus effecting actions; and *connector neurones*, which are mainly found within the central nervous system and relay information to and from several neurones. See also *neurotransmitter*.

Neurosis A broad category of psychological disturbances believed not to have an organic origin and which are not *psychoses*. The major neuroses are *depression, hysteria, obsessions* and *phobias*. Usually the sufferer maintains contact with reality, recognizing that the symptom is irrational but still unable to modify it. The term has gone through several meanings since the eighteenth century when it meant a disease of the nervous system. The usage indicates a belief about the source of the problem and today neuroses are expected to have psychological causes, whether in the remote or recent past or the present.

Neurotransmitter A chemical involved in *synaptic transmission*. There are many different chemicals which serve as neurotransmitters, of which the best known are *serotonin, acetylcholine, dopamine, noradrenaline, endorphin* and *enkephalin*.

Neurotransmitters are highly influential in subjective experience as well as in more general brain functioning, and many of the *psychoactive drugs* have their effect by either blocking the uptake of specific neurotransmitters or by preventing their dispersal and causing a build up of the substance within the *synaptic cleft*.

New paradigm research An alternative framework for research, based on *hermeneutic* principles – in other words, emphasizing the importance of social experience and social meaning. Traditional psychological methodology is seen as deterministic, tightly controlled and often artificial, resulting in socially meaningless information. In essence, new paradigm research involves a rethinking of the relationship between the psychologist and the person or persons who are the subjects of psychological enquiry. In new paradigm research, people are seen as active collaborators or participants in the study, whose opinions and experiences have value. This stands in direct contrast to conventional psychological methodology, which has tended to assume that its 'subjects' were there to be manipulated or tricked by the experimenter, and that good empirical investigation consists of 'controlling', or preventing, any human influences from individual subjects from affecting the research.

New paradigm research is therefore closely linked with the growing interest in *ethical* issues in psychology, and with the approach to social enquiry known as *ethogenics*. It tends to involve non-experimental methods of investigation, such as interviewing people and asking them about their experiences, or the dynamic real-world approaches exemplified in *action research*. It also tends to involve methods of analysis which are directly concerned with identifying the social meaning in the material, rather than with simple quantification. See also *qualitative analysis*.

Nicotine Sometimes described as one of the most addictive drugs ever known, nicotine is one of the most popular *recreational drugs* taken in industrial society. It is usually smoked, but sometimes chewed, and has been clearly implicated in lung and mouth cancers, and heart disease. Among its psychological effects, it includes a slight sedative effect. Nicotine is picked up in *receptor sites* in the motor end plate of the muscle fibres, thus reducing the uptake of *acetylcholine*. Consequently, nicotine withdrawal often leads to increasingly restless sensations, as the muscles become more receptive to acetylcholine; and to increased lability of the *autonomic nervous system*, accentuating both positive and negative emotional reactions.

Node of Ranvier The small spaces which occur in the *myelin sheath* along the *axon* of the *neurone*.

Noise Stimulation which does not carry any information. Often this will be a sound, but noise can occur in any sensory channel. Noise is of most interest when it accompanies information (which would tend to be called the signal) and therefore makes it more difficult to detect or interpret the signal accurately. A measure of detectability of a stimulus is to divide the strength of the signal by the amount of noise: the signal/noise ratio. The term is also used within psychology in its more usual sense of a strong auditory stimulus which may be of interest as a source of stress or of deafness.

Nominal scale The most basic way of attaching a value to an object or event, by classifying according to the category to which it belongs. Classifying psychological disorders as anxiety, depression, obsession, etc is an example. There is no arithmetic basis for putting nominally scaled items into any particular order and therefore very little that can be done statistically with such a scaling. However the frequencies in the different categories can be counted and then analysed, if appropriate, by chi square or at least by identifying the *mode*.

Nomothetic Concerned with the formation of general laws, usually of behaviour. Nomothetic principles are to do with that which is abstract, universal or generally applicable to humankind. See also *idiographic, hermeneutic*.

Non-contingent reinforcement *Reinforcement* which is not dependent on a particular action or response from the organism involved. Such reinforcement is often involved in the development of superstitious behaviour.

Non-directive therapy The group of therapies and counselling techniques that consistently avoid making value judgements about what the client has done, is doing, or should do. See *client-centred therapy*.

Non-parametric statistics Statistical techniques which do not require that the data should fit requirements such as interval scaling and *normal distribution*. Because they use less of the information in the data they are usually less powerful than *parametric statistics* but they are less likely to give misleading results. See *levels of measurement*.

Non-technological society A term used to describe societies which maintain their traditional economic systems and cultures, such as are found in some parts of Africa, Australia, and South America. In colonial times, such cultures were often referred to as primitive, but a deeper knowledge of them has shown that their levels of sophistication are extremely high, but centre around a more ecologically balanced style of living, rather than around technological development. Consequently, the term non-technological societies is increasingly used as providing a more accurate description.

Non-verbal communication (NVC) Communication through signals other than those used in language. For example *posture*, appearance, smell and a range of specific behaviours such as *pupil dilation, facial expression* and the pattern of *eye contact*. Non-verbal communication takes place through a number of different *non-verbal cues*, which can be combined in various ways. Some researchers have estimated NVC as being more than four times as powerful as verbal communication, though one could imagine that trying to teach the A level psychology syllabus non-verbally would be rather laborious. An understanding of the cues and use of non-verbal signals forms the basis of most *social skills training*.

Non-verbal cue A signal which conveys some kind of communication to an observer without involving the use of language. Non-verbal cues are usually considered to be of seven main types: *paralanguage, proxemics, posture, gesture, facial expression, eye-contact*, and dress. Some theorists additionally consider that *ritual* and ritual symbolism should also be regarded as an important medium of non-verbal communication.

Noradrenaline A *neurotransmitter* and *hormone* which is commonly involved in emotional reactions as a main transmitter of the *sympathetic division* of the *autonomic nervous system*, as well as within the brain itself.

Norepinephrine The American name for *noradrenaline*.

Norm The range of values within which the members of a particular population can be expected to function. *Psychometric tests* will list different sets of *standardized* norms for various groups of subjects, perhaps '45-year-old housewives' or '6-year-old boys in middle class homes'. Users of the tests can then compare their results with the norms of a comparable population. Within developmental psychology, norms are used to determine whether a child is performing so far out of the normal range as to require special treatment.

Normal curve The bell-shaped curve which is produced when data from a *normally distributed* population are plotted as a frequency distribution.

Normal distribution When Francis Galton began measuring a number of human characteristics he found that they could be plotted as a frequency distribution which consistently took the form of the *normal curve*. Many of the sources of data that psychologists deal with fit a normal distribution either because the population has that form or because the measure has been constructed deliberately to provide it, e.g. intelligence tests. The normal distribution has therefore been an important basis for many *parametric* statistical tests such as *t-tests* and *analysis of variance*. Because it is mathematically clearly defined it can easily be used to define aspects of a set of scores, particularly to indicate the probability or the implausibility of any specific score. Once the mean and the standard deviation of a normal curve are known, the frequency with which scores will be found a given distance away from the mean can be accurately computed. These frequencies are given as tables of z-scores in statistical texts. The values of most interest to psychology are those that will occur no more than 5% and no more than 1% of the time as these are the conventional levels which count as evidence against the *null hypothesis*. It is also a feature of the normal distribution that the *mean*, the *mode*, and the *median* have the same value. Problems with the normal distribution arise because many sets of data are not actually distributed precisely or even approximately, in this way. Yet the convenience of tests based on the distribution means that they are often used anyway, and we can usually only guess at how much influence this has on the results of the tests. Tests which do not assume a normal distribution are called *non-parametric tests*. [f.f.]

Normality A state which is usually considered to be unremarkable: the opposite of *abnormal*. In attempting to identify normal and abnormal behaviour for the purposes of psychological classification, three alternative approaches are often specified. Firstly, normality is taken to be behaviour which is accepted as usual, or as frequently occurring. Abnormal behaviour is then regarded as behaviour which is uncommon, or at least which is infrequently acknowledged. (In some cases, such as the imagined 'seeing' of a recently dead relative, the experience may actually be very common though not often openly acknowledged). A second definition of

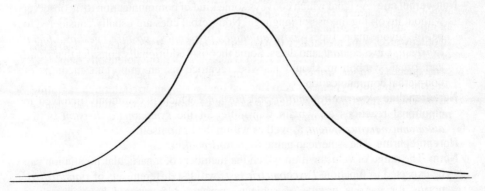

A normal distribution curve

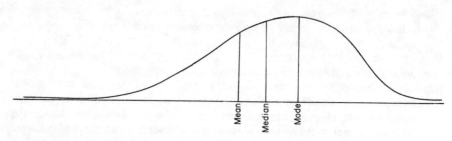

A normal distribution curve with a negative skew

normal behaviour, is behaviour which conforms to accepted norms, or social demands. In this event, social consensus becomes a major factor in decisions concerning normality and abnormality. The third approach concentrates on statistically common behaviour, irrespective of consensus. This approach rests on the assumptions of the Gaussian (*normal*) *distribution*. The problem here is that with this approach, people who are statistically uncommon in a highly valued direction (e.g. of extremely high IQ) are also defined as abnormal.

NREM sleep Non-rapid eye movement sleep: the stages of sleep in which *rapid eye movements* do not occur. See *orthodox sleep, sleep.*

Nuclear family A family consisting of two parents, one of each sex, their off-spring, and possibly the previous generation of parents and/or spouses of the offspring. The nuclear family has been treated as the basic family structure on which Western society is based, so people have been concerned to discover that it is much less common than had been supposed. In fact there are grounds for supposing that it never was as common as had been assumed.

Nucleus A dense area within the cell body, which contains structures necessary to the life and development of the cell, including *chromosomes* and messenger *ribonucleic acid.*

Null hypothesis A prediction in a research study that the outcome of the study could have been simply a consequence of chance factors, and not really of the experimental conditions. The null hypothesis can never be totally ruled out, which is why it is wrong to make statements like 'the t-test is non-significant so the null hypothesis is true'. Instead, the amount of confidence which can be placed in the results of a study is expressed in terms of how low is the probability (p) that the null hypothesis is correct. This is known as the *significance level*, or *confidence level*. In most student experiments, the acceptable level of significance is set at $p < 0.05$. In other words, if the probability that the null hypothesis is correct is less than 0.05, or 1/20, the results will be accepted. For research with more potentially damaging consequences, more stringent significance levels are used. Another way of describing the null hypothesis is to say that the results occurred through sampling error. This is referring to whether the sample of subjects in the study accurately represents its parent *population*. See *normal distribution, statistics.*

NVC See *non-verbal communication.*

O

Obedience Within social psychology, obedience has been studied as the social phenomenon which enables an individual to perform actions when instructed to by someone else, which they would not consider when acting independently. Largely initiated by the work of Milgram (1973), the study of obedience shows how the *demand characteristics* of a situation appear to enable the individual to suspend their own conscience, and to perceive themselves as having had no option but to do so. See also *autonomous state*.

Obesity Excessive weight. Obesity is usually defined in terms of body weight being a certain percentage above the ideal weight for that person's age, sex and height. The percentage varies but is often either 15% or 30%. This vagueness is not crucial as there is no absolute standard for 'ideal weight', which is largely a cultural judgement.

Object concept The idea that objects have a continuing existence, whether the individual is paying attention to them or not. Although this has been disputed by philosophers, the operational concept is an important one for the young child to develop in its interactions with the world; and the way in which this happens has been extensively studied as part of *cognitive development*.

Object constancy The perceptual process by which adjustment is made to the fact that objects have a continuing existence even when not being attended to. See also *object concept, size constancy, shape constancy*.

Object permanence See *object concept*.

Object relations theory A *psychoanalytic* theory developed primarily by Melanie Klein and W. Ronald Fairbairn in Britain as a reaction against Freud's concentration on instincts. Objects are the people, parts of people or things to whom the individual relates. Infants are believed to relate only to separate parts of people, such as the mother's breast. The ability to perceive the parts as belonging to a whole person, with both their good and bad aspects, has to be learned. Only a whole person can be recognized as having their own feelings, needs, etc. which ought to be respected, so only a whole person can be the object of a mature relationship. Psychological disturbance in adults is believed to result from problems in object relations in childhood, with the more severe conditions reflecting problems earlier in development, hence the emphasis by Klein on the breast as the first, crucial, part object. Therapy is directed to resolving the relationship with bad or persecutory objects internalized by the patient so that they can make mature relationships with people and not just use them as vehicles for their own gratification.

Objective test A test which can be marked without any need for subjective judgements. For example, multiple-choice tests and intelligence tests are regarded as objective by most psychologists.

Observational learning *Learning* which occurs as a result of observing the behaviour of others. As such, observational learning includes the two processes of *imitation* and *identification*, and is an important component of *social learning theory*.

Observational study A study which involves watching what happens, in a given context, rather than intervening and causing changes. Observational studies may take place in a variety of conditions, ranging from a highly controlled laboratory setting to uncontrolled 'field' conditions. Similarly, the observation itself may be undertaken in a number of ways, ranging from the use of electronic equipment, to the presence of a human observer, to the active participation of the observer in the interaction under study. No matter how well controlled they are, observational studies can only provide *correlational* data as without the direct manipulation of variables, such as occurs in an *experiment*, causality cannot be inferred. The major methodological problem is that the presence of the observer, particularly if filming is used, is likely to influence the behaviour being observed. See *participant observation*.

Obsession An idea or image that persistently enters thought despite being unwanted and recognized as abnormal. See *compulsion*.

Obsessive-compulsive disorder A neurotic disorder in which the person is unable to resist spending a lot of time in *obsessional* thoughts which are usually absurd and obscene, and carrying out pointless rituals – *compulsions*. The condition is extremely distressing and associated with a high level of anxiety. Psychoanalysis regards it as a personality disorder in which tremendous efforts have been made to suppress and control emotions, with the obsessions and compulsions being the denied aspects of the self which are breaking through the defences. At an extreme the person may spend so much time on the ritual thoughts and acts that they are unable to do anything else at all. Behavioural approaches view this as an outcome of conditioning processes in which the ritual is reinforced because it provides temporary relief from the anxiety of tackling some real task.

Occam's razor A scientific principle which states that, given a choice between two possible solutions or theoretical explanations for a given problem, the simpler one of the two should be adopted. Also known as the law of parsimony.

Occipital lobe The lobe of the brain which is found at the back of the head. The occipital lobe contains the *visual cortex* of the cerebrum. See also *parietal lobe, frontal lobe, temporal lobe*.

Occupational psychology The use of psychological knowledge and principles in the study of people at work, or in any productive occupation. Occupational psychology and *industrial psychology* are closely linked, but occupational psychology has a wider range than just the study of people in industrial situations, as it includes such occupations as that of housewife, novelist, and unemployed person. See *organizational psychology*.

Ocular dominance columns Arrangements of cells in the visual cortex of the brain, identified by the Nobel prize-winners Hubel and Wiesel. They found that cells dealing with the same elements of visual stimulation (see *simple, complex* and *hypercomplex cells*) were arranged in columns running perpendicularly to the surface of the brain, and that these columns alternated in a highly regular fashion between those receiving visual information from the right eye and those receiving information from the left eye. It is thought that this arrangement helps the brain to compare the different images from the two eyes – using binocular *disparity* as a depth cue. [f.]

Oedipus complex In Freudian theory, a process occurring during the *phallic stage* (around 3 to 5 years) in which the child wishes to possess the parent of the opposite sex and so sees the same-sex parent as a rival. As this parent is also powerful and successful, the child will feel threatened, but also tends to resolve the conflict by *identifying* with the rival parent. Neo-Freudians, particularly of the *object relations* school, have shifted the emphasis onto earlier relationships with the mother, so that

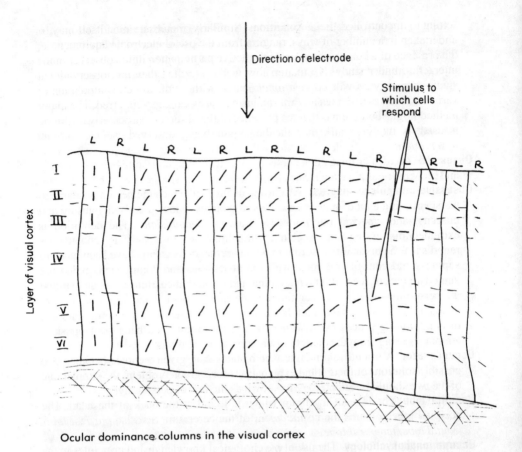

Ocular dominance columns in the visual cortex

oedipal conflicts have come to be seen either as occurring at a younger age, or as less important as a source of psychological disturbance. The oedipal process is regarded as applying just as much to girls as to boys.

Olfactory cortex A strip of the cerebral cortex which runs along the base of each *temporal lobe*, and receives information from the scent receptors in the nose. This area is concerned with the analysis and interpretation of smells.

One-trial learning A very rapid form of learning, through *classical conditioning*, in which just one experience or event is sufficient for a lasting learned association to occur. Most examples of one-trial learning have to do with food or pain, and are thus regarded as linked to very basic survival mechanisms. If consumption of a specific food is followed by vomiting, or if contact with a specific stimulus is followed by a painful experience, then a strong avoidance behaviour will result which is highly resistant to *extinction*. The forms of one-trial learning that are specific to the species and which seem to have a biological basis are examples of *prepared learning*. One-trial learning has also been associated with instances of *superstitious learning*.

Ontogeny The origins and development of the individual. The most well-known use of the word is probably in the phrase 'ontogeny recapitulates *phylogeny*', which is a biological principle popular at the beginning of the twentieth century, stating that the stages of growth of each individual member of a species mirrored the evolutionary development of the species itself. So, for instance, much was made of the idea that the human foetus in its early stages had structures which resembled gills, a tail, etc. Although this idea is now regarded as contentious, or even dubious, it was highly influential: for example, Piaget's study of *cognitive development* in the child was undertaken because of his interest in the evolution of abstract thinking and formal logic. By looking at how children developed logical processes, he hoped to identify the evolutionary stages by which rational thought had evolved.

Open system A system which is open to receive energy or information. Open systems are therefore able to develop, and will tolerate new structures within them, as opposed to a closed system.

Operant Any unit of behaviour which has an effect (of any kind) on the environment. Also known as operant behaviour, it is the basis of the conditioning of voluntary behaviour. Unless behaviour which has some kind of effect in the environment is produced spontaneously, the *law of effect* cannot come into play, and the behaviour will continue to be emitted more or less randomly.

Operant conditioning A process of stimulus–response learning of voluntary behaviour, which occurs as a result of the consequence of actions produced by an organism (animal or human being). The idea is that the learning of an appropriate action or operant is likely to be reinforced (strengthened) if the action is followed by a pleasant consequence (see *law of effect*). This increases the probability that the action will occur again. *Reinforcement* in operant conditioning can be positive or negative. If positive, the action is directly rewarded; if negative, it is indirectly rewarded by the removal or avoidance of something unpleasant. The other major class of conditioning is called *classical conditioning*. See also *primary* and *secondary reinforcement, partial reinforcement, reinforcement schedules.*

Operant strength This is a term used to describe how strongly a response acquired through operant conditioning has been learned. There are two main measures of operant strength: *resistance to extinction* and *response rate.*

Operational definition A definition which identifies something by its effects. An operational definition may not form an ideal definitive statement, expressing all aspects of the topic being defined; but it needs to be good enough to allow some *empirical* investigation of the topic. For instance, systematic work on *sustained attention* became possible only when researchers adopted the operational definition of attention as being the detection of relatively small changes in stimuli from within a varied background, e.g. picking out one particular signal on a radar screen. Failures to detect the target stimuli were accepted as evidence of failure to attend. Although this was not an ideal definition of attention itself, it served as a useful operational definition. Apart from giving clear rules by which the phenomenon can be identified, the definition also has to be close enough to the accepted meaning to be acceptable to most researchers. See also *signal detection task.*

Operations Manipulations of objects or concepts. The major use within psychology is in Piaget's theory which is largely about the different kinds of *cognitive* operations, particularly logical manipulations, which are carried out by children at different ages. See *concrete operations* and *formal operations.*

Opiates Drugs which have both *analgesic* (pain relieving) and *narcotic* (sleep inducing) effects. Opiates include naturally occurring drugs such as opium and morphine, drugs synthesized from the natural substances, such as heroin and some synthesized chemicals which have the same properties. There are also several naturally occurring opiates, of which the most well known are the *endorphins* and *enkephalins*, which act as *neurotransmitters* in the brain. Some foodstuffs, e.g. milk, are thought to contain small traces of naturally occurring opiates. Opiates are widely used both as clinical and as recreational drugs, and in general are highly *addictive*.

Opponent processing A theory originally proposed by Hering as an explanation of *negative aftereffects* – especially those to do with colour. Hering located opponent processing as occurring in the rod and cone cells of the retina, but more recent research indicates that it takes place in the second retinal layer of bipolar neurones. The idea is that cells, or groups of cells, have two different and complementary modes of operation. One group of cells responds to red stimuli when in one mode, and to green stimuli when in the other; a second group responds to blue or yellow stimuli; and a third group responds to light or dark. Over-stimulation of any one system through continuous presentation of just one of the paired stimuli results in compensation when the stimulation stops: the opposite stimulus is experienced as the cells gradually return to normal functioning.

Optic To do with the eye and vision.

Optic chiasma A point within the brain where the optic nerves from each eye meet. At this point, nerve fibres carrying messages from the left side of each *retina* combine, and pass on the left side of the *thalamus* and then to the *visual cortex* on the left hemisphere. Those carrying messages from the right side of each retina combine and pass on to the right side of the brain. This cross-over and recombination of nerve fibres is thought to be instrumental in *depth perception*, specifically in the process of *binocular disparity*, in which the image from each eye is compared.

Oral To do with the mouth. Oral functioning is particularly important for the young infant and seems to be a major source of pleasure. The oral region is quite mature in good time for birth, and the newborn is able to coordinate sucking, swallowing and breathing. In Freudian theory the oral is the first of the psychosexual stages, and fixation at the *oral stage* is said to produce adult tendencies such as greed, mania and depression, and a tendency to engage in oral behaviours such as smoking and lecturing.

Order effect An experimental effect which arises as a result of the order in which two tasks are presented. Order effects are of two main kinds: *practice effects*, where the subject becomes more skilled at a given task as a result of practice, and so performs better in later conditions of the experiment; and *fatigue effects*, where the subject becomes tired or bored as the study progresses, and so performs worse in later experimental conditions. See also *counterbalancing*, ABBA design.

Ordinal scale A system of measurement in which the basic units can be ranked. See *rank, levels of measurement*.

Ordinate The vertical or Y-axis on a graph. By convention this axis usually carries the measure of outcome, or dependent variable. See *abscissa*. [f].

Orectic To do with desire or appetite. The term is only likely to be encountered as an opposite of *cognitive*.

Organ of corti The structure in the *inner ear* which effects the *transduction* of vibration to electrical impulses, which are then transmitted to the brain for interpretation. The organ of Corti consists of two membranes, the basilar membrane and the tectorial membrane. Between the two are hair cells, which trigger off an electrical impulse when vibrated. This then passes to the fibres of the

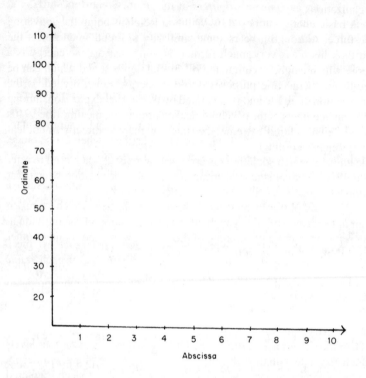

The coordinates of a graph

auditory nerve, which are embedded in the basilar membrane. See also *place theory, frequency theory*.

Organic disorder A disorder which is known, or at least believed, to be due to a physiological or organic malfunction. The extent to which psychological disorders result directly from organic brain dysfunction is one of the major controversies in the field. Psychiatrists are more likely than psychologists to believe that disorders are organic. The term is sometimes used as a contrast, for example to distinguish organic *psychoses* from those that are better understood psychologically, and are called '*functional*' psychoses.

Organism A term used during the *behaviourist* era of psychology to describe animals and human beings, when talking in terms of simple (stimulus–response) learning processes or motivational states, such as hunger or thirst drives. The use of the term in preference to animal or person was intended to signify: (1) the way that stimulus–response learning applied to all active creatures alike, as the basic building block of behaviour; and (2) the dispassionate objectivity of the scientist, whereby people were to be regarded simply as units which emitted behaviour, irrespective of sentimental human values.

Organizational psychology The study of how people act and interact in organizations. Although sometimes regarded as a part of *occupational psychology*, organizational psychology is increasingly accepted as an area of study in its own right. It is distinguished from *industrial psychology* in that it includes public sector and voluntary organizations.

Orgone energy A basic energy proposed by Wilhelm Reich as being the activating universal life-force. Although it bears some similarity to Freud's concept of the *libido*, Reich took his ideas very much further, arguing that orgone energy is a physical energy which can be accumulated by special devices, and which can be utilized directly for therapeutic purposes. Orgone energy, Reich argued, is the source and motivation of all life and is generated by free sexual expression, among other things. Many members of the psychological community at the time (the 1940s to 1950s) found Reich's claims extreme; the state saw them as directly fraudulent and prosecuted Reich accordingly.

Orientation The angle at which something is arranged, or exists. When used to refer to an individual's theoretical stance, it means the attitude or position which that individual adopts towards a specific theory or school of thought.

Orienting reflex A set of physiological and behavioural changes which occur in response to an unexpected stimulus which attracts the attention of the individual. The orienting reflex includes a positioning of the body towards the sound or other stimulus, keeping the body very still, a dilation of the blood vessels in the head, *EEG* changes and alterations to muscle tone, heart rate and breathing. This combination of physiological changes means that the individual is more prepared to receive the stimulus. The opposite pattern, when a stimulus is being excluded, is called the defensive reflex.

Orthodox sleep Ordinary, quiescent sleep which does not involve *rapid eye movements* (REM), or the experience of *dreaming*. Orthodox sleep occurs at four levels or stages, which correlate with the subjective experience of being lightly or deeply asleep, and which each show characteristic *EEG* patterns. Stage 1 sleep is entered first, and is the lightest form of sleep with a fairly regular EEG pattern. Some dreaming may take place during this stage. The sleeper then progresses through the stages to the deepest level of stage 4 in which the EEG is very irregular with large spikes. In this stage it is very difficult to awaken the sleeper and in children, bedwetting, night terrors and sleep-walking may occur. The pattern changes through the period of sleeping – see *sleep cycles*. Orthodox sleep is also called NREM (non-rapid eye movement) sleep. See also *paradoxical* sleep.

Osmoreceptors Although not empirically discovered, osmoreceptors are thought to be receptors in the brain which respond to changes in fluid composition in brain cells, and so are thought to act as signals to the brain for the experience of thirst.

Otoliths Small particles of a bony substance which float in the fluid-filled semicircular canals of the *inner ear*. The canals are lined with hair cells, which produce an electrical impulse when the otoliths come into contact with them. In this way, movement and turbulence of the fluid in the semicircular canals is detected, which is an important factor in the sense of balance.

Outer ear The part of the ear which is in direct contact with the outside world. It includes the pinna and the lobe (the two external parts of the ear itself) and the auditory canal. The pinna and lobe serve to direct sound waves into the auditory canal, and they pass along it in the form of waves of changing air pressure until they come into contact with the ear drum, or tympanic membrane. This marks the boundary between the outer ear and the middle ear, and vibrates in response to the air pressure. See also *middle ear, inner ear*.

Over-compensation An excessive response in attempting to overcome a disadvantage or difficulty. There is usually an implication that the person who does this is abnormally affected by the original problem. For example a short person who goes to exceptional lengths to disguise or compensate for their height would be judged to be excessively sensitive about it. The term is more often used as a derogatory term in lay language than in psychology, where it has no technical meaning. See *compensation*.

Over-extension The tendency, found particularly in young children acquiring a language, to apply words too widely, e.g. calling all men 'Daddy'.

P

Paedophilia A condition in which an adult is sexually attracted to children and can only achieve sexual arousal with them. See *sexual abuse*.

Paired-associate learning A learning task which involves the association, or linking together, of two stimuli, usually words. This form of learning task was extremely popular in the study of *memory* throughout the 1950s and 1960s, but of recent years has been heavily criticized for its artificiality.

Pairing Presenting two stimuli in such a way that they always occur together.

Pandemonium model A *hierarchical* model of cognition, first proposed in the late 1960s, which forms an interesting example of *bottom-up* processing. It has been mainly concerned with feature-recognition in perception, and the way in which the identification of features can be combined to result in meaningful percepts. The model proposes a hierarchical organization of 'sub-demons', 'cognitive demons', and 'decision demons'. There are myriads of sub-demons, each of which is tuned in to detecting specific aspects of a stimulus, such as specific letters in a word. When a stimulus occurs, the appropriate sub-demon shrieks. The more similar the stimulus is to the demon's template, the louder it shrieks. The decision demon at the next level in the hierarchy is faced with the task of deciding which of the shrieking sub-demons best represents the stimulus, taking into account other shrieking sub-demons responding to subsequent stimuli (hence the name of pandemonium model). As the overall picture becomes more complex, general cognitive demons come into action, which operate at a higher level, and represent complete concepts or *schemata*. Because of the idea of competition between the demons at each level, this model is well able to cope with the explanation of our response to ambiguous stimuli; but some consider it to be weak in explaining some of the more general aspects of active cognition.

Panic attack An *anxiety* disorder in which the person experiences sudden and unpredictable attacks of acute anxiety or terror which have no organic cause nor are they a response to any threat in the environment. The anxiety is increased by the fact that the person does not know when another attack will happen, and cannot make any sense out of what is happening to them.

Paradox A situation in which two or more rules combine to give an impossible outcome, like the Cretan who said 'all Cretans are liars'. Paradoxes have been much studied in logic and mathematics but for psychologists the chief interest is in those that trap people into apparently crazy behaviour. See *double bind* for an example. Some therapists believe that many symptoms result from paradoxes in the person's life and so are best treated with a 'counterparadox' designed to free them. A common example would be to instruct the person to have their 'uncontrollable' symptom at a particular time. If they have the symptom then it shows they can control it. If they do not have the symptom it shows the symptom can be prevented, i.e. it is controllable. As with any other powerful therapeutic technique, paradoxical injunctions can be ineffective and potentially harmful unless they are used with respect and sympathetic understanding for the patient.

Paradoxical sleep A name given to the type of sleep in which *rapid eye movements* occur (it is also called REM sleep), during which *dreaming* occurs. It was named paradoxical in the 1960s, as a result of the discovery that EEG patterns shown in this type of sleep suggested that the sleeper was only lightly sleeping and would wake easily; whereas in reality they proved very difficult to wake by some stimuli (like loud noises) but easy to waken by more meaningful events like having their name spoken. See also *sleep cycles; orthodox sleep*.

Paralanguage The *non-verbal cues* which are used during speech, and include speech sounds, such as 'er' and 'um', the timing of utterances or inflections and accents. Paralanguage is an important part of communication through speech, but provides information independently of the actual verbal aspects of the communication. A measure of the importance of paralanguage to speech is the way that in written language, punctuation is needed to substitute for the additional information normally added through tones of voice or pauses.

Parallel play A form of *play* in which two or more children play alongside each other without direct interaction. It is common in young children before *social play* becomes possible.

Parallel distributed processing (PDP) A *computer simulation* system which works on the principle that human reasoning often involves the simultaneous operation of more than one sequence of argument. Consequently, PDP involves simulation programmes which operate several different logic chains simultaneously, with considerable cross-linking between them. The particular value of this approach appears to be that it is capable of producing novel or unexpected outcomes in computer problem-solving. The general approach to computer simulation expressed by this technique is also known as *connectionism*.

Parallel processing The processing of *information* in such a way that more than one set of *operations* is happening simultaneously. Models of parallel processing were introduced to *cognitive psychology* in an attempt to account for the extremely rapid ways in which people can search for information, taking several features into account apparently all at the same time.

Parameter A mathematical measure of some characteristic of a population, such as the mean. The same measure in a sample is called a *statistic*.

Parametric statistics Statistical techniques that have been developed on the assumption that the data are of a certain type. In particular the measure should be an *equal interval scale*, and the scores should be drawn from a *normal distribution*. Also, different samples should be independent of each other: the choice of items or scores in one sample should not have affected the choice of items or scores in another. Because construction of the tests is based on these assumptions, using them on data which do not fit the assumptions can give misleading results, though there seem to be no clear answers about how serious a problem this is. Parametric statistics are usually preferred because, by using more of the information available in the data, they are more powerful in detecting significant effects. The alternative is to use non-parametric tests which do not make the same assumptions about the data. However apart from the loss of power, there are not always non-parametric versions of techniques like analysis of variance and factor analysis. See *levels of measurement*.

Paranoia A disorder in which the person is dominated by thoughts of persecution, grandeur, and sometimes jealousy. Intellectual functioning is not impaired and great ingenuity may be shown in interpreting every event to fit with the paranoid belief.

Paranoid schizophrenia A *schizophrenic* condition of which the main feature is paranoia, but the consistency of the beliefs found in paranoia is missing.

Parapsychology The study of phenomena which resemble psychological events but which are not explainable by any accepted principles. Parapsychology claims such processes as *telepathy, psychokinesis, clairvoyance* and *extrasensory perception*. All of these have the characteristic that they cannot be reliably reproduced under controlled conditions and are therefore very difficult to study scientifically. In practice the main disagreement is over whether these phenomena ever happen under any circumstances. The term parapsychology is accurate in that it means 'beyond psychology' but whether that will turn out to mean that it goes further, or just that it is outside the scope of psychology is a matter of opinion at present.

Parasympathetic division The division of the *autonomic nervous system* which comes into action during the quiescent emotions, such as contentment or sorrow. The parasympathetic division is also concerned with processes for restoring and conserving bodily resources such as digestion; and storing glycogen and other reserves which have been depleted by the action of the *sympathetic division* of the autonomic nervous system.

Parenting A term used instead of *mothering* either to emphasize that any adult could be providing the care, or to refer to a specific aspect of care of the young that is undertaken by either parent.

Parietal lobe The large area of the cerebrum located behind the *central fissure* and above the *occipital lobe*. See also *frontal lobe, temporal lobe*.

Parkinson's disease A progressive neural disease which affects older people, producing gradual loss of motor control, noticeable trembling of the limbs, and resulting eventually in paralysis. Parkinson's disease is known to be caused by *dopamine* deficiency in the brain, and the symptoms can sometimes be alleviated by treatment with a drug known as L-dopa, which converts into dopamine in the brain itself. Unfortunately, this form of treatment also has distressing side-effects, so it is not seen as a fully satisfactory method of managing the disease. Long-term use of many anti-psychotic drugs can produce a set of symptoms similar to Parkinson's disease, known as drug-induced Parkinsonism.

Partial reinforcement Reinforcement in an *operant conditioning* process which is not given every time the desired behaviour is shown, but only some of the time. This is also known as *intermittent reinforcement*, and produces a somewhat slower but stronger form of learning which is more resistant to *extinction*. See also *reinforcement schedules*.

Participant observation A research technique in which the researcher takes a full role in the group being studied, often without the knowledge of the other members. In this way the distortion produced by the presence of an observer is minimized, and the researcher can obtain a fuller appreciation of the experiences of the group. See *observational study*.

Pattern perception The way in which different perceptual features of shapes or figures are recognized as belonging together and forming a pattern of stimuli, rather than being separate and discrete. Without pattern perception, our subjective experience would be simply of patches of light and dark or of patches of colour without any linking of the stimuli into meaningful units. The basis of pattern perception is *figure-ground organization* the inherent tendency for our perceptual system to organize sensory data into meaningful figures set against backgrounds. This organizational principle results in pattern perception, and is evident in the perception of other sensory modes, such as music or speech perception, which, involve pattern perception in linking and distinguishing the different components of the information.

Pavlovian conditioning See *classical conditioning*.

PDP See *parallel distributed processing*.

Peak experience The rare experience of feeling for a moment complete and at one with oneself and the world. *Maslow* regarded peak experiences as important, but not essential, aspects of *self-actualization*.

Pearson's product–moment correlation (r) A measure of *correlation* which uses *interval data*. It is thus a *parametric* test and makes the standard assumptions about the data. It is the preferred measure of correlation if the data are suitable. If not, then *Spearman's rank correlation* may be used. See *statistics*.

Pecking order An idea taken from the observation that chickens seem to have a social hierarchy in which anyone can peck those below them but not those above. The unfortunate character at the bottom is under attack from all of the others, and is literally 'hen-pecked'. The term has been extended to describe any social heirarchy in which there is a clear and specific definition of the order in which people or animals are dominant, and which is more technically called a *dominance hierarchy*.

Peer group A group composed of people from similar backgrounds and of equal status. Most commonly used to indicate that the group is composed of children of equal age.

Penis envy In *psychoanalytic* theory, the envy that girls are claimed to feel about the fact that boys have a penis and they do not. Freud believed that women experience penis envy throughout their lives but this is now a deeply unfashionable point of view for which Freud has received his fair share of *interpretations*.

Percept The impression which the person receives of that which is being perceived. The percept is the subjective or internal experience which represents an object or event in the external world.

Perception The process by which we analyse and make sense out of incoming sensory information. Perception has been studied extensively by psychologists, and now forms part of *cognitive psychology*. Perception can be distinguished from sensation, which concerns the stimulation of sensory receptors and may also be restricted to the earlier stages of processing incoming information. However there is no fully agreed definition, and some theorists such as Ulric Neisser regard perception as identical to the rest of cognition and so would make little or no distinction between the two. Perception includes several distinct areas, such as *visual perception; person perception; auditory perception;* and the perception of other forms of information such as *pain, gustatory, tactile* or *olfactory* input.

Perceptual constancy The way that a person's perception adjusts itself so that the world is seen as constant, despite the changes in stimulation produced at the sense organs. The perceptual constancies enable us to perceive events more accurately in terms of their meaning, for example people are seen as the same size however far away they are. There are many forms of perceptual constancy, of which the most studied have been *size constancy, shape constancy, colour constancy*, and *location constancy.*

Perceptual defence The idea that the perceptual system is more likely to receive information which is non-threatening, and has higher *thresholds* for perceiving information which is psychologically threatening to the individual, meaning that such information is less likely to be detected or recognized. See also *defence mechanisms.*

Perceptual set A state of readiness or preparedness to perceive certain kinds of information rather than other kinds. Perceptual set is a powerful phenomenon, which links closely with *selective attention* and which can be affected by a range of circumstances, such as prior experience, emotion, motivation, culture and habit.

Performance A term used in experimental psychology for the level at which a person or animal performs on a particular task.

Peripheral nervous system A term for those parts of the nervous system which are not included in the *central nervous system* (the brain and spinal cord). The peripheral nervous system accordingly includes the *autonomic nervous system* and the *somatic nervous system*, composed of *motor* and *sensory neurones* carrying information to and from the CNS.

Person perception The application of methods for studying and understanding perception to the perception of people. Person perception is fundamental in the process of understanding other people, and often, by implication, ourselves. It has been found to have the usual features of perception when it is operating in conditions in which the object is complex and the conditions are difficult. That is, it is highly influenced by set and expectations, and by the needs, fears and wishes of the observer. Person perception is an active and highly researched area within psychology, involving the study of *attribution*; of *non-verbal communication*; of interpersonal *attitudes*; and of social *memory.*

Personal attributions *Attributions* which are seen to apply just because that particular person was involved. They therefore tend to relate to some unique or identifying characteristic of that person. For example, passing a very high-level music exam on the cello would be likely to be attributed to the special characteristic of exceptionally high talent. If the attributed causal sequence would have happened whoever was involved, it is classed as 'universal'. Some writers such as Seligman treat personal attributions as the same as *internal attributions*. See also *consensus.*

Personal constructs A unique set of ideas about the world and the people in it, which each individual develops and uses to make sense of the world, and to function effectively in it. Personal constructs were proposed by George Kelly (1955) as the individual theories which people use to generate hypotheses in order to explain their experience. Kelly's model of the person was of 'man-as-scientist' – that the person was actively making sense of the world by formulating hypotheses about it, and then testing them, much as a scientist investigates their chosen subject area. By identifying the special, personal set of constructs which the individual uses, a therapist would be far better placed to understand that person and to assist them with their problems in living. Kelly's was thus an *idiographic* theory, concerned with the uniqueness of the individual and how he understood his world. The form

Personal space

of assessment known as the *repertory grid*, which Kelly developed, allows the therapist to utilize the individual's own constructs in analysing their experience.

Personal space The distance which people keep between themselves and others during everyday activities. The distance will vary depending on the individual's culture, on the circumstances, and on their relationship with the other person: we tend to position ourselves more closely to intimate friends than we do to strangers. Personal space is a manifestation of *proxemics*, and an important *non-verbal cue*; it is often described in terms of *territoriality*. [f.]

Personalism The degree to which the actions of others are perceived as directed particularly towards yourself. There is evidence that we tend to overestimate the extent to which this happens, that is, we over-personalize. See also *attribution, hedonic relevance*.

Personality Those relatively enduring features of an individual which account for their characteristic ways of behaving. We put this forward as a useful definition but many alternatives would be possible. The differences are not a matter of accuracy but of deciding which approach to the subject is most likely to be productive. Some uses of the term 'personality' refer to patterns of behaviour rather than their causes, or, more narrowly, to the social *roles* that a person adopts. Some theories are concerned with the way the structures underlying personality are formed (Freud, for example), and in general the *psychodynamic* approaches stress personality as an integrated whole, more than the sum of its parts (see *personality dynamics*).

Other theories are trying to attain a biological basis, for example Eysenck's theory of *types*. Another approach is to measure different aspects of people on the assumption that their behaviour is the product of many *traits*. Completely different is the line taken by Walter Mischel who claims that there is little evidence of stable structures within people that cause them to behave in certain ways. Instead he suggests that, as far as human behaviour is consistent at all, it is consistent because people tend to spend their time in particular kinds of environment and so behave in recognizable ways. Mischel would claim there is no such thing as personality as defined above.

Personality assessment A system for measuring the personality characteristics of different people. Personality assessments may utilize the format of objective testing, as in a *personality inventory*, or of *projective tests* such as the *Thematic Apperception Test* or the *Rorschach* ink-blot test; or they may be *phenomenological* tests such as the *repertory grid* or the *Q-sort*.

Personality disorder A term for the very broad class of psychological disorders which seem to arise from long-term characteristics of the person. Roughly, the term applies to conditions which reflect what the person is, rather than how they behave (*behaviour disorders*). Examples include *psychopathy*, and *paranoia*.

Personality dynamics An approach to understanding behaviour in terms of the active interplay of aspects of the personality structure. Freud's account of personality in terms of interactions between the *id, ego*, and *super-ego* is the classic example.

Personality inventory A personality test which takes the form of a set of straightforward questions about the individual's behaviour; which is used to build up a *personality profile* or to assess personality traits quantitatively according to a predetermined set of criteria. See *trait theory*.

Personality profile A system for describing the outcome of a personality test which assesses the individual in terms of predefined *traits*. Rather than just providing a single score as the outcome of the test, an image of how the individual has scored on each of the set of traits is given, usually graphically.

Phallic stage The third *psychosexual stage* in Freudian theory, in which the child's interest focuses on the penis. Having based a significant part of personality development on something possessed by only half of the species, Freud's theory ran into all kinds of complications, and some accusations of male chauvinism, about this stage. The phallic stage ends with the *oedipal conflict*, and is generally concerned with issues of potency. The term 'phallic' is used when the emphasis is on symbolic aspects of the penis.

Phantom limb The experience, by people who have had a limb amputated, of sensations as if they still had the limb. Of interest to psychologists because it is informative about how the *body image* is maintained.

Phenomenal field A term used by perceptual theorists to describe the totality, or complete picture, of what is being perceived.

Phenomenology The study of phenomena, or events, as perceived by the individual. A psychologist who adopts a phenomenological approach will emphasize the importance of perceiving events as the people involved in them see them; an objective perception of events is seen as being of very limited value when it comes to understanding human behaviour.

Phenotype The developed organism which results from the interaction of the genetic characteristics which were inherited from the parents, and the environment in which development occurs. Although the term carries an idea of an end-product, the phenotype is a dynamic rather than a static phenomenon, which, as both genetic and environmental influence continue throughout life, is constantly developing and changing. See also *genotype*.

Pheromones Chemicals which are released into the atmosphere from the body and which provide a form of communication, as they are detected by other members of the species. Many species release distinctive pheromones to signal sexual receptiveness, and synthesized pheromones are often used by animal breeders to facilitate mating of their animals. Although pheromone detection appears to be linked to the sense of smell, it is not identical to it, as many pheromones seem to exert a direct effect on hormone balance.

Phi phenomenon An illusion of movement brought about by the sequencing in illumination of adjacent lights. If one light comes on when the other goes off, and the light next to it goes on when that goes off, what is perceived (assuming it happens reasonably quickly) is an impression of one light moving across from the location of the first one to the location of the last. This phenomenon is widely used in illuminated advertising signs, and can sometimes be very convincing. Should the lights be arranged in a circle, the perceived circular motion is seen as describing a circle of smaller diameter than the actual arrangement of the lights. It is thought that the phi phenomenon is a manifestation of the *Gestalt* psychologists' *principle of closure* occurring with dynamic stimuli rather than with static ones.

Phobia A *neurotic* disorder in which there is a stong and persistent fear of objects or situations which is not justified by any danger that they pose. The sufferer will be aware that the fear is irrational but will make strenuous attempts to avoid the feared situation. Often, the symptoms can best be seen as attempts to avoid the (very unpleasant) sensations of anxiety, rather than being closely tied to the feared object. Phobias may be attached to a wide range of situations, and particular forms are indicated by putting the appropriate term (usually in Latin or Greek form) in front of the word, as in *agoraphobia* and *claustrophobia*. Specific phobias can usually be treated effectively by behaviour *therapy*, but many, like agoraphobia, incorporate a fear of social interaction, and are more difficult to treat.

Phobic disorder The standard term to cover all of the *phobias*.

Phobic reaction The full range of behaviours shown by a person suffering from a *phobia*.

Phoneme A basic unit of spoken language: a speech sound. Phonemes are not the same as syllables: a one-syllable word, like 'cat' for instance, is made up of three distinct phonemes, which are combined to produce the syllable, or *morpheme*.

Phonemics The study of regularities and distinctive patterns in the combination of phonemes in spoken language.

Photoreceptors Cells in the *retina* which respond to light and so are necessary for vision.

Phylogenetic scale An approximate scale which attempts to chart an evolutionary progression through different types and groups of species, to human beings. Species are ranked in order of approximate similarity to humans, with primates being closest and thus seen as higher up the phylogenetic scale, and with fish and reptiles being seen as significantly lower down. The concept of the phylogenetic scale is an inherently misleading one, implying as it does that evolution proceeds in a linear fashion, and that other species can be seen as steps towards an ultimate goal; but the concept of species similarity which it contains is sometimes useful in evaluating studies in *comparative psychology*. If we want to generalize to human behaviour, it makes more sense, to take examples from other primate groups, or at least mammals, than it does to take them from species which are far less closely related, such as birds, insects or fish.

Phylogeny The evolutionary processes by which a species develops its characteristics.

Physical punishment Punishment which involves some identifiable material consequence, such as keeping a child in after school, or loss of pocket money. Although corporal punishment is included in this category, the term physical punishment is used to describe a wider range of punishments than simply physical chastisement. Compare *psychological punishment*.

Physiological correlate A physical change which accompanies a behavioural or psychological response. The term is used to avoid making assumptions about causality. It may be recognized, for instance, that a cognitive event such as concentration or sleep is accompanied by physiological changes in the body. However, the relationship between the physiological change and the event itself is not a simplistic causal one, and so the term physiological correlates is adopted as a description.

Physiological need Identified by Maslow as being the lowest level in his *hierarchy of needs*; physiological needs are the requirements for physical functioning, such as the needs for food, water, etc.

Physiological psychology The study of the way in which human behaviour and cognition are influenced or performed by processes which take place physically within the body. The term 'physiological' is preferred to 'biological' because such influences are usually exerted by whole systems of physical functioning operating together, such as is demonstrated in the *fight or flight response*, or the sensory information processing systems. Physiological psychology is often seen as being inherently *reductionist* as it explains behaviour in terms of the actions of neurones and chemicals, but many physiological psychologists maintain an *interactionist* approach to the subject, in which physiological factors are seen as contributing to or influencing behaviour, but not necessarily determining it.

Pilomotor response The response of the hair of the body standing on end at times of extreme fear or rage. In many animals, this forms an impressive signal, resulting in

The pilomotor response

the animal looking much larger and, presumably, more fearsome to a would-be attacker. It is also sometimes used to fluff-up the hair to provide added protection from cold. In human beings, owing to the shortness and near-invisibility of much body hair, the pilomotor response simply results in the skin appearance known as goose-pimples, as the contraction of the small muscle at the base of each hair pulls the surrounding skin into a small bump. [f.]

Pineal gland A gland situated centrally in the brain, which was once thought to be the seat of the soul. The pineal gland is known to be involved in the hormonal changes which signal the onset of puberty, and some recent research has indicated that it may also be involved in *diurnal* and seasonal hormonal variation, although the precise functioning of the gland is far from being known.

Pitch The term used to refer to the frequency of a sound, in terms of the perceived highness or lowness of the note. See *place theory, frequency theory*.

Pituitary gland This is the main, or 'master control' gland of the endocrine system. The pituitary gland has a direct link with the *hypothalamus*, and secretes *hormones* which carry signals to all the other glands, stimulating their operations.

Pivot words Words that children seem to use in the earliest stage of *language* acquisition, as a base to which a large number of other words (called open words) can be attached, e.g. 'allgone car', 'allgone Daddy', 'nasty allgone'. Pivot words were once thought to be the basis of *grammar* and it was hoped to extend the concept to utterances of three and more words. The idea is no longer widely used in theories of language development.

PK See *psychokinesis*.

Place theory The idea that the pitch or frequency of a sound is identified by the brain in terms of the specific region of the *organ of Corti* which is stimulated, with high tones triggering off the hair cells nearest to the oval window, while lower tones stimulate hair cells further along the cochlea.

Placebo A fake or dummy form of medication which is given during experimental trials investigating the effects of drugs. A placebo resembles the drug under study, but has no measureable effect on the body. By comparing the results of those who have had the drug and those who have had the placebo, experimental effects such as the influence of subjects' beliefs are controlled. In most such studies, a *double-blind control* will be used, such that *experimenter effects* are also controlled, as the experimenter is not aware who has taken the placebo and who has taken the drug.

Play There is no satisfactory definition of play. Either it is defined by exclusion, amounting to saying it is not work, or the definition makes assumptions that fail to capture the appropriate range of activities. Such a situation is usually an indication that there is no adequate theory. Our ignorance about play comes under two headings: functions and cause. Function is concerned with the role that play has in the development of the individual, and how it came to be present in the species. Theories here concentrate on the fact that much play results in the development of skills that will be useful later in life, but that play is uncoupled from serious consequences and so can be indulged in safely by the immature organism. The issue of cause, whether a particular child will play in a particular situation, is even less well understood, with most work having been done under the heading of *exploration*. Clues to both function and cause can be found by studying the forms that play takes. Most of this research has concentrated on pre-school children as many of their activities involve play. See *fantasy play*.

Play therapy A range of techniques in the diagnosis and treatment of children which exploit the child's tendency to play. Often materials such as puppets, dolls, or just a piece of string may be provided and kept for the child between sessions. In play the child will explore concerns which cannot be expressed in words, and the therapist both learns about the child's problems and can help the child to find ways of dealing with anxieties and difficulties.

Pleasure principle In Freudian theory, the basic function of the *id* is to pursue pleasure. In infancy, with a high degree of dependency on caregivers, and before the ego with its *reality principle* has developed, pleasure must be achieved either through dependency on caretakers or through fantasy. In this context Freud wrote of pleasure as the reduction of tension, as if all stimulation or arousal at least for the infant is unpleasant.

Pluralistic ignorance This occurs when everyone in a group believes something but no-one expresses it and so each person thinks they are alone in their belief. Cases of *bystander apathy* and crowd behaviour may depend at least partly on pluralistic ignorance combined with *conformity* to presumed beliefs of the rest of the group. The concept also informs the idea of a 'silent majority'.

Point of subjective equality (PSE) The value of a continuously variable stimulus at which it appears to be identical to a standard stimulus. Not usually measureable directly but derived by a variety of *psychophysical* techniques. For example, judgements may be obtained from a subject about whether a series of lines are larger or smaller than the standard line, and the point at which they switch from larger to smaller is called the point of subjective equality.

Polygenic Resulting from the action of many genes. The genetic element in overall body height, for instance, results from the action of several genes contributing to the development of different parts of the body. *Phenotypic* characteristics which are polygenic will show continuous variation in the population, as height does.

Polygraph A device used to measure *autonomic* arousal which takes measurements of a number of different indices, and provides a multiple read-out ('poly' is from Greek, many). Typically, a polygraph will take measurements of blood pressure and *heart rate, EEG, galvanic skin resistance* and muscular tension. By such means, it is possible to tell when an individual is under stress, and so polygraphs are often used as 'lie-detectors'. A considerable amount of controversy surrounds their use in criminal investigation, as it is not possible to distinguish the stress produced by telling lies from that produced by other factors, e.g. anxiety on behalf of someone else, or physical pain.

Pons A region in the lower part of the brain which serves to connect the two halves of the *cerebellum*, and may be involved also in mediating *dreaming* sleep.

Population All of the cases within a given definition, for example, all of the women in Britain; all of the schools in Huddersfield; or all of the people in a given lab class. Psychological research is nearly always only able to take a *sample* from a large population, though researchers will often want to generalize their results to the whole of the population.

Positive regard Liking, affection or love for another person. The term was used by Carl Rogers to describe what he considered to be one of the two basic needs of the human being: the need for positive regard from others. This, he thought, could be conditional upon appropriate behaviour or unconditional, but as a basic need, it would have to be satisfied. Rogers's form of therapy requires that the therapist provides the client with *unconditional positive regard*. See also *self-actualization*.

Positive reinforcement *Reinforcement* which provides something that the organism wants, likes or needs – a reward of some kind. See also *negative reinforcement*.

Positivism A belief that reliable information can only be obtained about events that can be observed directly. It therefore claims that science should only deal with observables and not *hypothetical constructs*. Behaviourism in its more primitive forms has been the clearest example of a positivistic approach within psychology. An even more restrictive version, called logical positivism, claims that a hypothesis can only be regarded as scientific if there is a way in which it can potentially be disproved by empirical observation. Logical positivism has been largely abandoned or superseded, but it was always more popular among philosophers of science than among psychologists, who mostly just got on with the job of studying hypothesized psychological processes such as motivation.

Post-hypnotic amnesia The forgetting of information as a result of a suggestion made while the subject was under *hypnosis*, and which occurs after the hypnotic state has finished. Post-hypnotic amnesia is often described by subjects as feeling like 'tip-of-the-tongue' forgetting and can often last for several days.

Post-hypnotic suggestion A suggestion made to someone while they are in a hypnotic state, which concerns behaviour which they will undertake once the hypnotic fugue is over. In the case of relatively trivial forms of behaviour, this is often performed by the subject, who typically says that they 'just felt like doing it'. Post-hypnotic suggestion has sometimes been presented by Hollywood film makers as being so powerful that it could force a subject to act against their will, but this represents part of the Hollywood mythology of hypnotism, which bears little resemblance to the real thing. It is not possible to force someone to do anything against their will, either during hypnosis or through post-hypnotic suggestion: the state of hypnosis itself necessarily involves the willing cooperation of the subject throughout.

Postpartum depression Depression in mothers within a few months of the birth of their babies. To be distinguished from 'the blues' which is very common around

Postural echo

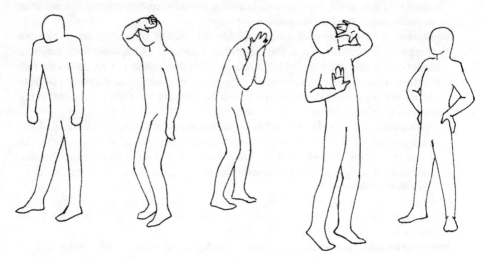

Posture can be a rich source of non-verbal communication

the third day after the birth but which is not depression and does not persist. Some evidence is beginning to emerge that suggests that depression in women is no more common following birth than it is in other women of the same age. If this turns out to be the case, then there will be little reason to suppose that postpartum depression is in any way caused by the pregnancy or birth.

Post-traumatic amnesia *Amnesia* which results from some kind of accident, such as that resulting from a blow to the head or from severe brain damage.

Post-traumatic stress disorder (PTSD) A set of symptoms commonly found following any kind of extremely disturbing experience. Recent research on concentration-camp victims indicates that the disorder may persist over many years. It is being found to be a common response in victims of rape, political torture, and natural disasters such as the Bradford stadium fire and the Zeebrugge ferry sinking.

Postural echo A *non-verbal* signal which often indicates friendliness or that two people are in substantial agreement. While the participants are engaged in a social exchange (such as a conversation) they may be seen to be adopting (usually unconsciously) the same posture; or mirroring each other's posture if they are face-to-face. It may be used consciously by therapists and salesmen to produce a feeling of rapport in the client. [f.]

Posture A powerful *non-verbal* cue which is commonly used to indicate attitudes or emotions. It is about the positioning of the body, the relative arrangement of the limbs. Posture is commonly, though usually unconsciously, taken as a communicative signal, and may make a considerable difference to how a verbal message is understood. See also *postural echo, non-verbal communication*. [f.]

Power When applied to a statistical test, this refers to the ability of the test to identify an effect, or reject the null hypothesis, when an effect is present. In any test the power increases as the sample size is increased, but some tests are intrinsically more powerful than others. In general, tests which use more of the information in the data are more powerful. So a *t-test*, which calculates the amounts by which scores differ, is more powerful than a sign test, which just uses information about whether scores are bigger or smaller. *Parametric tests* are more powerful than *non-parametric tests* for this reason.

Power law A law propounded by S. S. Stevens which states that the subjective strength of a stimulus is equal to the physical strength of the stimulus raised to a power (squared, cubed, etc). Like *Fechner's law*, the power law relates to the fact that as a stimulus becomes stronger, bigger changes are required in order to have the same psychological effect. The power law differs from Fechner's law in the mathematical expression of the relationship.

Practice effect An experimental effect in which apparent changes in the *dependent variable* happen as a result of the subject gaining practice in the task during the course of the experiment, and therefore improving their performance. Practice effects are usually controlled by *counterbalancing* the order of presentation of the conditions of the study.

Pragmatics An approach to studying *language* which concentrates on the functions that language performs rather than on the structure of the language itself (*linguistics*).

Precocial animals Animals which can move about as soon, or almost as soon, as they are born or hatched. Research into *imprinting* has traditionally centred around work with precocial animals as they show the phenomenon in a clear and unambiguous form.

Precognition A knowledge of future events which is not based on judgement but on direct and certain perception of them. As a branch of *parapsychology*, precognition requires more than the certainty that dinner will be provided this evening, and implies a special form of knowledge, different from any that is understood by psychologists. Anyone who could operate precognition reliably would presumably be either very rich or very depressed.

Preconscious Thoughts and knowledge that are not at present in the conscious, but which are not *repressed* and so can be brought into consciousness at will. Freud proposed that the preconscious lies between the *unconscious* and the conscious mind, which it more closely resembles.

Predictive validity See *validity*.

Prejudice Literally 'pre-judgement', prejudice refers to the maintenance of a prior attitude irrespective of new or contradictory information. It is commonly used in connection with negative or discriminatory social attitudes, such as *racism* or *sexism*; it may also refer to a pre-determined favourable judgement by which the individual ignores relevant negative information. See *stereotype*.

Prelinguistic thought The forms that thinking takes in children before they have developed language abilities. Knowing about prelinguistic thought may help us to understand the extent to which adult thought may be independent of language.

Premature Used of babies who are born before *gestation* is complete. New obstetric techniques mean that premature babies now survive from a much earlier stage of development than was previously possible. Psychologists are concerned about the effects on the immature baby of being exposed to intense environmental stimulation at a time when the nervous system is biologically adapted to the protective environment of the womb.

Pre-moral stage The first of Kohlberg's three stages of *moral development*, in which moral judgements are seen entirely instrumentally, in terms of whether or not the individual is likely to be detected and/or punished.

Pre-operational stage The second of *Piaget's* stages of cognitive development.

During this *stage* children are unable to think in terms of logical concepts such as *conservation* or *reversibility* and they are dominated by perceptual features of their world. The stage starts from about 2 years of age, at the end of the *sensori-motor stage*, when *object permanence* is first seen. It ends at about 7 years when the child starts the stage of *concrete operations*.

Prepared learning The finding that to some extent, organisms are biologically prepared to learn certain *associations* very easily. The commonest example is that animals who experience nausea will associate this sensation with whatever they last ate rather than other kinds of stimuli, even if these were more intense and more recent. It is sometimes called the 'Garcia effect' after its discoverer, but also known as the '*sauce Béarnaise* phenomenon' after an account by Martin Seligman of an experience of being sick, due to a stomach virus, after eating a steak with his favourite sauce, and being unable to face eating it ever again. In fact the effects can be overcome and Seligman has had a lot of free meals while people have tested its permanence.

Presenting symptom A patient will usually come into therapy on the basis of a complaint about a particular symptom. This is called the presenting symptom, a reminder that some other symptom may be the real problem which may only emerge in the course of therapy. A major dispute about evaluating different forms of therapy is based on the issue of whether resolving the presenting symptom counts as a success. Many behaviourists will work only on the presenting symptom and end therapy once it is eliminated. *Psychodynamic* therapists are more likely to see the presenting symptom as a kind of ticket that enables the patient to get into therapy so they can then start dealing with the real problem.

Primacy effect An effect of the presentation of stimuli, whereby those items which are presented first tend to be recalled more readily than those which are presented later on. Primacy effects do not only occur with simple memory tasks, but have their counterparts in person perception, too, whereby those characteristics of a person which are first encountered tend to be applied more readily than any characteristics which emerge or are learned later. In *memory* studies the primacy effect is part of the *serial position effect*.

Primary abilities The fundamental mental abilities which were suggested by Thurstone as forming the basis of *intelligence*. There were considered to be seven of these: memory, verbal ability, word fluency, number, spatial awareness, perceptual discrimination and reasoning.

Primary process In Freudian theory, the more primitive kind of mental process which is present in the functioning of the *id* from birth. It is seen as the way the *unconscious* operates later in life, being governed by the *pleasure principle* and not following the same laws as conscious or *secondary process* thinking. For example, primary processes take no account of time and space, so unconscious memories of frightening childhood events are just as real, powerful, and present as current perceptions.

Primary reinforcement A *reinforcement* which satisfies a basic need or drive in the organism. See *operant conditioning, secondary reinforcement*.

Primary sexual characteristics Those signs of someone's gender which are directly to do with reproduction. These include the genitalia: the penis, and testicles in the man and the vagina and clitoris in the woman. See also *secondary sexual characteristics*.

Principle of closure Probably the most powerful of the *Gestalt* principles of *perceptual organization*, the principle of closure refers to the perceptual tendency towards complete forms and shapes. So a set of disconnected lines is likely to be seen as indicating an incomplete shape if this is at all possible, rather than simply being taken as independent stimuli. The principle of closure also extends into the perception of movement, in the form of *stroboscopic motion* and the *phi phenomenon.*

Principle of parsimony See *Occam's razor.*

Principle of proximity One of the *Gestalt* principles of perceptual organization, which states that stimuli which occur close to one another will tend to be perceived as grouped together, all other things being equal.

Principle of similarity One of the *Gestalt* principles of *perceptual organization*, which states that similar stimuli will tend to be perceived as grouped together, all other things being equal.

Privation A lack, throughout development, of some requirement. Privation should be distinguished from deprivation, in which the requirement was available for a period and then removed. Experiments in which animals are raised with no contact with a mother are privation studies, but they are often referred to as *maternal deprivation.*

Proactive interference When information which as already been learned interferes with the learning of new material. Proactive interference is particularly common when someone is trying to learn a set of similar tasks within a relatively short period of time. It may account for the *primacy effect.*

Probabilistic concept A concept which involves a set of characteristics which its members are likely to share, but need not necessarily do so. For example: a chair is a probabilistic concept, in that one of its distinctive features is that chairs usually have four legs. But there are many styles of chair which don't fit into that category: it is probable, but not necessary. In practice, most concepts used by human beings are probabilistic in nature. See also *classical concept.*

Probability The likelihood that an event will occur. Formally, the probability is calculated by dividing the number of ways the event could occur by the number of all possible events. So the probability of getting a red apple out of a barrel on a single trial is given by the number of red apples divided by the total number of fruits (red apples + green apples + oranges, say) in the barrel. The probability of getting a red apple ranges from 0 (no red apples, so zero probability of getting one) to 1 (nothing but red apples so you are certain to get one). Much of statistics within psychology amounts to assessing the probability of an obtained result, given certain assumptions. If the probability is very low, then the assumptions are unlikely to be valid. See also *null hypothesis, statistical significance, binominal distribution.*

Probe A stimulus such as a word or a digit which is used to explore something, e.g. how much information is retained in *short-term memory.* In this example, the subject would hear a sequence of digits read out, and would then be told a specific number and asked if that had been in the list. The number would be acting as the probe.

Problem-solving The study of the various strategies used by people, computers, and sometimes animals, to achieve solutions, usually of highly specified puzzles. By having the problem clearly specified it is hoped that the detailed cognitive processes involved in problem-solving will become apparent. It is, however, not clear whether the findings from such research have application to more complex human problems, such as how to pass an exam or pay the mortgage. See also

*functional fixedness, Einstellung, Stroop effect, groupthink, brainstorming, crea-
tivity.*

Product–moment correlation See *Pearson's product–moment correlation.*

Programmed learning A technique for applying operant conditioning to classroom
learning. The information is broken down into small units, and presented to the
student in such a way that one unit leads naturally on to the next. Each unit involves
some kind of simple test question. If the student gets it right, they move onto the
next stage; if they get it wrong, they go back over the relevant material again. The
idea is that this approach maximizes *positive reinforcement* (in the shape of correct
answers) for the student, thus maximizing interest in and application to the learning
process. As an example of pure operant *conditioning*, programmed learning has
been criticized on the grounds that knowledge of results is a cognitive rather than
a behavioural reinforcement. In classroom practice, the absence of social
interaction between student and teacher has often presented its own difficulties, and
programmed learning has tended to be introduced in a manner that is far more
limited than was previously envisaged.

Projection One of the ego-defence mechanisms identified by Freud, which involves
the individual attributing their own *unconscious* motives and ideas to another
person, or to an ambiguous situation. For example, a person who has not come to
terms with their own sexual drives might come to believe that many other people
engage in 'bad' sexual practices. Like other *defence mechanisms*, this is an
unconscious process, but it is often a useful signal to a therapist of things which
particularly concern the client.

Projective test A form of personality assessment which involves presenting the
subject with ambiguous stimuli, and requiring them to indicate how they would
interpret each stimulus. The idea is that the reply will indicate some of the concerns
of the individual's *unconscious* mind: themes and events which particularly
concern them at a subconscious level will be *projected* onto the ambiguous
material. Well-known examples of projective tests are the *Rorschach ink-blot test*
and the *thematic apperception test.*

Proprioception The perception of the positioning of the limbs, and of movement.
Proprioceptors are sensory neurones which convey information from joints and
muscles to the *central nervous system*. Proprioception is commonly considered to
represent a sixth basic sense; dealing with internal rather than external sensory
information. See also *kinaesthetic.*

Prosocial behaviour The opposite of antisocial behaviour; prosocial is used to refer
to behaviour which involves helping others or making a positive gesture towards
them in some way. It is commonly used in discussions of *bystander intervention*
and *altruistic behaviour.*

Protocol A plan of the steps or stages involved in the solution of a problem. See
algorithm.

Proxemics The study of *personal space* and the use of touch as non-verbal cues in
communication.

Proximo–distal A sequence of development identified by Gesell in early studies of
infant development of motor co-ordination and incorporated into his theory of
maturation. Gesell observed that motor control seemed to be acquired over the
more central regions of the body first, and only later did the extremities (hands,
feet) become co-ordinated. From this he argued that development proceeded in an

orderly direction, which he called proximo–distal ('from near to far'). See *cephalo-caudal*.

PSE See *point of subjective equality*.

Psi The ability to perform paranormal tasks. See *parapsychology*.

Psyche The mind. Psychology was originally defined as the study of the mind.

Psychedelic drugs Drugs which induce altered *states of consciousness*, commonly resulting in a heightened awareness of colour and perceptual imagery. Psychedelic drugs have been used as *recreational drugs* for centuries, but were named 'psychedelic' during the 1960s, as a result of their association with a particularly vivid form of visual art, involving massed swirling of colours and similar imagery. Drugs classified as psychedelic include the *hallucinogens: mescaline, LSD*, and *psilocybin*.

Psychiatry The medical treatment of abnormal behaviour, or of mental disturbance. Psychiatrists are always medically qualified and therefore, unlike clinical psychologists, may prescribe drugs. Approaches used by psychiatrists range from a concentration on physical methods of treatment through to *psychotherapeutic* approaches in which they may work in very similar ways to some clinical psychologists. Within the National Health Service psychiatrists have a statutory responsibility to deal with all of the cases sent to them. Psychologists do not have this requirement and may therefore be able to spend more time on fewer patients.

Psychic determinism The idea that all psychological processes have been directly caused by something, whether that be unconscious motives or fixations, as suggested by the *psychoanalystic* theorists; or the patterns of firing of nerve cells in the brain brought about by conditioned responses to stimuli.

Psychoactive drugs Drugs which affect psychological experience, such as moods, consciousness, or awareness. Although this is a very general term, it is most often applied to the groups of drugs commonly used for psychological purposes, such as *anti-anxiety drugs, sedatives, tranquillizers, anti-depressants, stimulants*, and *hallucinogens*.

Psychoanalysis The method of psychological treatment originated by *Freud* and developed by various of his followers, the *neo-Freudians*. The major features of psychoanalysis are the use of *free association* to uncover *defences* which may then be *interpreted* by the psychoanalyst in order to bring *unconscious* material into the conscious. More recently the roles of *transference* and *counter-transference* have been emphasized. The term is not applied to the methods of those such as Carl Jung who broke away from Freud and whose approach is called *analytic psychology*.

Psychoanalytic theory The theory of personality development and human functioning developed by Freud. Psychoanalytic theory was continually elaborated and refined by Freud during his lifetime, and the process has continued since his death, so there is no single 'psychoanalytic theory'. However, basic to these models are the ideas of the *unconscious* with its effects on everyday behaviour, the *psychosexual stages* of development, and the personality structure of *id, ego*, and *superego*. Freud gave the theory a strong biological flavour, and the assumption that adult behaviour is powerfully influenced by childhood experiences remains fundamental. One major development of the theory has come from the *object relations* theorists who have emphasized the importance of experiences from very early infancy. Psychoanalytic theory has had an extremely wide influence on Western culture, for example in the understanding of art and literature. Many of Freud's original insights are now regarded as common sense. The theory has been attacked as unscientific, particularly by Carl Popper, on the grounds that it does not make claims which can be disproved. It has also been attacked on the grounds that several of its claims have been empirically disproved.

Psychobiology See *biospsychology*.

Psychodrama A set of therapeutic techniques, orginally introduced by Jacob Moreno in 1925, in which people are helped to act out troublesome emotions or situations. Regarding the situation as a play, and helped by the therapist and, usually, other group members, the patient can come to understand the problem better, and can try out alternative ways of responding in safety.

Psychodynamic All of the theories of human functioning which are based on the interplay of drives and other forces within the person. *Psychoanalytic theory* is the clearest example and the term psychodynamic is often used to refer specifically to this class of theories.

Psychogenic Having a psychological origin or cause. Used particularly of disorders for which no *organic* cause can be identified so it can be assumed, by default, that the cause is psychological.

Psychokinesis (PK) Bringing about a physical effect at a distance by psychological or more accurately *parapsychological* means.

Psycholinguistics The study of psychological aspects of language and the relationships between *language* and other psychological processes. Psycholinguistics deals with such questions as the interdependence of language and thought, language acquisition and the ways in which social experience and language acquisition interact, reading and *pragmatics*. Psycholinguistics is therefore a much broader field than *linguistics*, which is concerned with the origins and form of language.

Psychological assistant A recently introduced career grade in the U.K., to allow psychology graduates to work within clinical psychology departments. The grade is intended to replace that of *psychology technician*.

Psychological punishment Punishment which does not necessarily involve an explicit penalty, but which is more concerned with the communication of social expectation and the disappointment/sadness of people for whom the miscreant cares. Psychological punishment typically requires some form of act of atonement, such as an apology or an attempt to right the wrong. Some research has indicated that psychological punishment is more influential than *physical punishment* in changing behaviour and in producing strong consciences in children.

Psychological technician A person employed, usually in a clinical psychology department, to undertake routine psychological tasks such as the scoring of *psychometric tests* or the organizing of therapeutic events. This type of job is often undertaken by psychology graduates seeking relevant experience before undertaking postgraduate training in *clinical psychology*. The position is gradually being replaced by that of the *psychological assistant*.

Psychology Psychology has been defined in various ways, depending on the inclinations of researchers at the time that the definition was formulated. It has been variously defined as 'the study of the mind', 'the study of behaviour', 'the study of human experience' and 'the study of mental life'. It is difficult to produce a definition which will satisfy everyone, though we can state that it involves the study of human and animal behaviour and experience, examined from a number of different viewpoints and using a variety of techniques; most of which emphasize the importance of empirical evidence in support of explanatory theory. The field of psychology is divided, often somewhat arbitrarily, into different areas each of which has their own style. Some qualifiers of the term such as *developmental, social,* and *comparative,* refer to particular kinds of subject matter, with newer areas like *environmental psychology* being added as the field expands. Other titles: *clinical, educational, occupational,* refer to the psychological professions, while the term *applied psychology* refers to a general orientation which cuts across the whole field.

Psychometrics The measurement (usually through questionnaires or inventories) of psychological characteristics: 'mental testing'. Psychometric tests include *intelligence tests, personality tests, creativity tests,* and a whole range of tests used for personnel selection and *vocational guidance.*

Psychomotor retardation Slowing down of speech and movement found in severe depression.

Psychopathic personality A form of personality disorder in which the person lacks anxiety and guilt, disregards society's laws and conventions, and has no concern for other people. They may also be impulsive and aggressive. The condition does not fit readily into psychological classifications, but the term is extensively used in legal situations and allows certain kinds of offender to be treated in special hospitals. Also called 'antisocial personality disorder'.

Psychopharmacology The study of the psychological effects of drugs. See *psychoactive drugs.*

Psychophysics The study of the relationship between the experience of physical stimuli and the physical stimuli themselves. For instance, the study of the relationship between perceived levels of sound, and levels of sound as measured by physical instruments. See *Fechner's law, power law.*

Psychophysiology Sometimes used with the same meaning as *physiological psychology,* but usually having a more restricted meaning. In its restricted use the term refers to studies which use non-intrusive methods of monitoring physiological processes, e.g. surface electrodes, to provide information about psychological processes. See *physiological correlates.*

Psychophysiological disorder See *psychosomatic illnesses.*

Psychosexual stages In Freud's theory of personality development, the progression of bodily aspects through which pleasure is sought and towards which biological drives are directed. The stages are: *oral, anal, phallic, latency,* and *genital.* As with any *stage theory* each stage must be completed more or less satisfactorily in order for the next to be tackled. Failure to complete a stage means that a significant part of the person's resources will remain invested in that primitive source of gratification, and their personality will show relevant tendencies throughout adult life.

Psychosis A term to cover the most severe mental disorders such as *schizophrenia, bipolar disorder,* and *paranoid psychosis.* The person in a psychotic state loses contact with reality (see *reality testing*), has severe disturbances of thought and emotion which are not open to being changed by contrary evidence, and has little or no insight into their condition. Compare *neurosis.*

Psychosocial stages The term given to the eight life-stages proposed by Erikson. Each stage involves a basic conflict which the individual needs to resolve and which in turn provides a foundation for the later stages. In brief, the eight conflicts are: (1) trust/mistrust; (2) autonomy/doubt; (3) initiative/guilt; (4) industry/inferiority; (5) identity/role confusion; (6) intimacy/isolation; (7) generativity/stagnation; (8) integrity/despair. These basic conflicts arise at different stages throughout the individual's life, right up to old age and present the individual with a set of age-specific challenges to deal with.

Psychosomatic illness An illness which has its cause in psychological factors. Although the symptoms and discomfort of psychosomatic illness are genuine, and often highly distressing to the patient, the illness itself does not originate from a physical disorder of the body, but from some kind of mental disturbance or discomfort, often *unconscious* in nature.

Psychosurgery The use of surgical intervention in the brain to control behaviour. The most well-known form of psychosurgery is the operation known as lobotomy, the removal of the *frontal lobe* of the brain in order to induce quiescent behaviour in highly agitated, aggressive or psychotic individuals. A similar operation, *leuco-tomy*, involves the severing of the connections between the frontal lobe and the rest of the brain, leaving it in place and producing similar effects. Although largely discredited as a technique by neuropsychologists, in some areas psychosurgery is still performed to control *psychotic behaviour* – it is one of the more highly contentious aspects of *psychiatry*, and one which is not accepted by many psychiatrists.

Psychotherapy Usually the term covers the whole range of psychologically based treatments by which trained practitioners help people who have psychological problems. Sometimes the term is used in a more restricted way, most usually to refer to forms of treatment in which a psychotherapist and a single client tackle the client's problems through talking. Specific forms of psychotherapy may be identified by additional terms, for example *psychodynamic* psychotherapy covers forms of psychotherapy which have been based on one of the psychodynamic theories. Other forms are *non-directive therapy, cognitive psychotherapy* and *rational-emotive therapy*. See also *counselling, WEG*.

Psychotic behaviour Behaviour which is comparable to that shown by a person suffering a *psychosis*.

Psychotropic drugs This term literally refers to drugs which will promote or effect psychological growth. It was first used to describe the 'mind expanding' *hallucinogens* such as mescaline, psilocybin and LSD. The terms is in common misuse nowadays to refer to any drug which has an effect on mood, such as tranquillizers, which are more properly referred to as *psychoactive*.

Puberty The stage of physical growth during which the child becomes capable of reproduction. The occurrence of puberty is genetically controlled and so is a *maturational* process. In girls it is taken to start at the onset of menstruation, and in boys at the first presence of live sperm in the urine. As this sign is not readily visible, the growth of pubic hair is more commonly used. Although puberty is regarded as a period preceding *adolescence*, there is no clear definition of its end. It can be taken as lasting until the basic physiological structures required for reproduction have achieved a form recognizably similar to the adult state. Substantial psychological adjustments are required during puberty to cope with changing body shape and appearance, novel hormonal balances and associated emotional changes and changing sexual identity. Because the early stages of puberty are so visible, the substantial variations in age of onset (roughly from 10 to 14) can cause problems for both early and late developers. The average age of onset of puberty seems to have reduced by several years over the last century, which suggests that adjustments are being demanded at an earlier stage of psychological maturity.

Punishment The application of some kind of penalty or unpleasant event in order to suppress an unwanted form of behaviour. Although punishment is commonly used as a means of behavioural control, there is some evidence to suggest that it is of limited value by comparison with more directive approaches such as the direct rewarding of desired behaviour which occurs in *operant conditioning*. Note that punishment is not a form of *negative reinforcement*. See *physical punishment, psychological punishment*.

Pupil dilation The enlarging of the pupil of the eye. This happens mainly: (1) in darkness, or dim lighting, where the pupil enlarges so as to allow more effective vision; (2) under the influence of certain drugs, in particular *amphetamines* and *narcotic drugs* (see also *belladonna*); and (3) when the individual looks at someone or something which they like or are fond of. As such, pupil dilation is a very powerful *non-verbal cue*, indicating interpersonal attraction or empathy, and several studies have shown that people will respond more positively to others with dilated pupils (one reason for the low lighting common in many restaurants and night clubs).

Q

Q-sort A test often utilized in conjunction with *client-centred therapy*, to evaluate the individual's self-esteem in their own terms. The Q-sort consists of a set of cards, each of which provides a short statement about character or personality, which may be positive, neutral or negative. Clients are asked to sort the cards into piles which express how closely the statements fit with the individual's own *self-concept*: e.g. 'very like me', 'unlike me', etc. When all the cards have been sorted, the client is asked to sort them again, but this time in terms of their ideal self: 'myself as I would like to be'. The similarity or otherwise between the two sets of card-sorts provides a *correlation coefficient* indicative of the individual's self-esteem. Among other uses, the Q-sort has been employed in studies of the efficacy of client-centred therapy.

Qualitative analysis An approach to the analysis of psychological information which takes as its starting point the idea that the meaning of the information is the most important thing. Qualitative analysis is therefore not concerned with reducing psychological information to numerical data (quantitative analysis), but is concerned with identifying ways of extracting meaning in a systematic and reliable manner. See also *account analysis, thematic qualitative analysis, ethogenics, quantitative analysis*.

Qualitative difference A difference in kind, not simply in amount. If two things are qualitatively different, it implies that arithmetic comparisons between them are not appropriate, as they are of a different nature, like chalk and cheese. See also *quantitative difference*.

Qualitative research Techniques for obtaining psychological information which assume that the meaning of the information is the most important thing. The commonest techniques used in qualitative research are *interviews* and *case studies*. These methods allow greater freedom for the person who is the target of such research to determine what information is generated, so the quality and richness of information is greater, but at the cost of making *reliability* difficult to achieve. See also *hermeneutics, new paradigm research*.

Quantitative analysis An approach to psychological information which is primarily concerned with obtaining numerical information, which can then be analysed using statistics. Quantitative methods require the researchers to define the items to be measured in advance, and to control the situation so that only that information is recorded. The result is that high levels of *reliability* can be obtained and measured, but there are frequently problems with *validity*. Although traditional psychology has often tended to assume that only quantitative analysis is worthwhile, recently many psychologists have become increasingly concerned with the qualitative analysis of information, in addition to quantitative techniques. See also *psychometrics*.

Quantitative difference A difference in amount, rather than a difference of kind. See also *qualitative difference*.

Quota sampling A system of obtaining a *sample* for a study which involves identifying a set of representative sub-groups within the *population*, and taking a number of subjects from each of these sub-groups. The size of each sub-group in the sample depends on its proportional size in the original population. For instance, in a study of student attitudes to their Technical College, the sample would be picked to represent the same proportions of different types of students as were found in the college as a whole – if 10% of the students were on day-release courses, then 10% of the sample would be drawn from the day-release students. See also *sampling*.

R

Race differences Group differences between different races identified by use of *psychometric tests*. Because these tests usually measure something valued by European culture, and because their objectivity has been overestimated, findings of lower scores, for example on *intelligence tests*, of ethnic minority groups have been used as the basis for claims of racial superiority. These claims have then led to a rather more careful inspection of the evidence and it is now recognized that neither race nor intelligence can be defined or measured with enough accuracy to justify claims about the relationships between them.

Racism Discrimination, *prejudice* or unfair practice towards someone which occurs purely on the basis of their ethnic group or skin colour.

Random sampling A process of selecting a *sample* for an experiment or other empirical study, in such a way that any member of the *population* has an equally likely chance of being selected. Random sampling, when carried out appropriately, is considered to be the strongest sampling technique for avoiding bias in subject selection; if all members of the population have an equally likely chance of being selected, then if the sample is large enough it should reflect all the characteristics of its parent population.

Randomization A process of sorting subjects or experimental conditions into a random order so that no consistent pattern will be operating. For example if you recruit 20 volunteers from a class, the first 10 may differ in motivation or altruism from the last 10. It would be important to randomize the order of these subjects in an experiment rather than just put the first 10 to volunteer into the first condition and the rest into the second.

Range The difference between the highest and lowest of a set of scores. Range is the simplest and crudest *measure of dispersion*.

Rank To put a set of scores into order by size. The word can also mean the position of an item within a set of ranked scores. Ranking provides no information about how far apart adjacent scores may be, and so provides only ordinal data which must then be treated by *non-parametric* statistical techniques.

Rank correlation coefficient See *Spearman's rank–order correlation coefficient*.

Rapid eye movement (REM) sleep A form of sleep in which the body remains comatose, except for the eye muscles, which move rapidly and continuously. When woken from REM sleep subjects often report *dreaming*, and if an external stimulus, such as being lightly sprayed with cold water, is applied at this time, the dream content is likely to reflect the stimulus; in this case the subject might dream of being out in the rain. REM sleep occurs in phases throughout the night. Each phase usually lasts about 20 minutes, before the subject passes on to one of the deeper, quiescent levels of sleep. The phases become longer and more frequent during the course of sleep. Over the life span the time spent in REM sleep drops from about 8 hours in the newborn to about 1.5 hours in the elderly. The function of REM sleep is disputed, with theories ranging from those that see it as functional either in physiological restorative processes or as the phase in which the information acquired during the previous day is processed, to theories that it is left over from a previous stage of evolution. REM sleep is also known as *paradoxical sleep*. See also *sleep cycles*.

Rapport A feeling of psychological comfort in interaction with another person, based on feelings of trust and *empathy*. Used particularly about the relationship that is necessary between a *psychotherapist* and their client, or between a tester and their subject.

RAS See *reticular activating system*.

Rating scale A system of measurement, usually of attitudes, in which a person is asked to evaluate some stimulus material or idea on the basis of a predetermined scale which expresses degrees of liking or preference.

Ratio scale See *levels of measurement, equal interval scale*.

Rational-emotive therapy (RET) A form of *cognitive psychotherapy* developed by Albert Ellis. It is based on the idea that people make common logical errors, e.g. believing that it is necessary to be competent in every way, to have everyone love you, and to have whatever you want immediately. RET takes the form of persuading the client, by cognitive, emotional and behavioural means, to see things differently (correctly) so that their behaviour will be less destructive.

Rationalization Providing entirely rational and worthy explanations for one's behaviour which are designed to conceal from oneself, or from others, the less acceptable cause of the behaviour. As in 'we don't employ Blacks because the customers would not like it'. The process was identified by Freud as one of the major *defence mechanisms*, but it often takes the form of the basic *attributional error*.

Raven's progressive matrices An intelligence test which is designed to be *culture fair*. The test consists of a series of grids or matrices of 8 patterns from which the 9th can be deduced logically, and a set of patterns of which one is the missing 9th pattern and is therefore the correct answer. The special feature of the test is that it is entirely non-verbal and it is even possible to administer it to someone with whom the tester shares no language at all. Despite the attempt of Raven to make the test independent of culture, it still reflects some cultural assumptions and experience. Three examples of these assumptions are: (1) solving a puzzle whenever it is presented to you; (2) geometric shapes can be manipulated according to rules; and (3) familiarity with two-dimensional representation (line drawings). In many cultures manipulation of and/or interest in abstract forms of this kind are not regarded as particularly desirable human activities.

Reactance The tendency of people to be made uncomfortable by any restriction of their freedom of choice. Once such pressure is perceived people will often act in opposition to it.

Reaction formation A *defence mechanism* by which a person resists and denies an unacceptable motive by acting as if the opposite were true. The classic example of a reaction formation occurs in *homophobia*, in which the individual suppresses their own homosexual inclinations so strongly that they become extremely hostile to anyone expressing overt homosexuality.

Reaction time A measure of how quickly a person can produce an accurate response to a stimulus. Reaction time has been used by psychological researchers in a wide range of investigations, including *ageing, decision making, drug effects* and *vigilance*. It provides a rapid and reliable measure, which is highly sensitive to disturbance by additional or extraneous factors.

Reality anxiety In Freud's classification of anxiety he included those situations in which the anxiety is justified by a real external threat. See *moral anxiety*.

Reality principle In Freudian theory, the principle on which the *ego* operates. Whereas the *pleasure principle* is innate, the child has to learn about reality and how to operate in order to satisfy its needs. This developmental process is fundamental to the formation of the ego.

Reality testing A fundamental human tendency to check out one's understanding of the real world, particularly one's role in, and influence on, both physical and social reality. From infancy through early childhood there is a progressive development of the ability to distinguish between fantasy and reality. A failure to make the distinction in adulthood is taken as an indication of *psychosis*. *Personal construct theory* is largely concerned with the precise forms that reality testing takes.

Recall The first and strongest of the four forms of remembering identified by Ebbinghaus, recall refers to the retrieval of information on demand from memory storage. The other forms of remembering, in order, are: *recognition, reconstruction,* and *relearning savings*.

Recapitulation theory The now out-dated idea that individual development retraces the steps of the evolution of the species. See *ontogeny*.

Receiver-operating-characteristic curve (ROC curve) In *signal detection theory*, a graph in which the probability of hits and false alarms is plotted against the signal level.

Recency effect A learning effect in which the items which occurred most recently in a sequence are more likely to be recalled than those which occurred earlier on. See also *primacy effect*.

Receptor The term is usually used to mean sense receptor: a specialized cell or group of cells which picks up sensory information, either from within (see *propriocep-tors*) or outside of the body, and converts it into electrical impulses for transmission to the *central nervous system*. So, for example, the light-sensitive *rod* and *cone cells* of the eye are receptors, as are the hair cells in the *organ of Corti* in the ear, and the pressure-sensitive cells in the skin.

Receptor site A location on the dendrite of a neurone, opposite a *synaptic knob*, which is sensitive to and readily absorbs a specific chemical. The appropriate chemical is released into the *synaptic cleft* from vesicles on the synaptic knob of the opposing neurone, and functions as a *neurotransmitter*, rendering the receiving neurone more or less ready to fire. Receptor sites may also pick up chemicals with a similar structure, and many *psychoactive drugs* have their effect by being taken up at receptor sites appropriate for other chemicals: the *hallucinogens LSD* and psilocybin are picked up at receptor sites sensitive to the neurotransmitter *serotonin*, while *opiates* such as *heroin* and *morphine* are picked up at sites appropriate for the *enkephalins* and *endorphins*.

Recessive gene A gene which carries a developmental characteristic that only shows in the *phenotype* when the individual inherits a matching recessive gene on the other chromosome. If the paired gene, the *allele*, is of a different type and *dominant*, then the recessive gene will not influence that individual's development, although it could be passed on to children. Many common characteristics, such as red hair or blue eyes; and some genetic disorders, such as sickle-cell anaemia are carried on recessive genes; and so may appear to skip whole generations and appear in children of later generations.

Recidivism Repeated legal offences, such that the person concerned, the recidivist, appears in court on several occasions, not just once. A certain amount of work on *juvenile delinquency* reported by Rutter suggests that recidivism links very strongly with a continually stressful home life, at least for teenagers.

Reciprocal altruism Helping behaviour which occurs in a social context such that an individual, person or animal who receives help, in turn helps the individual who originally helped him. Reciprocal *altruism* often occurs over extended periods of time, and may not be recognized by a short-term ethological study.

Reciprocal liking The name given to a positive relationship between two or more people in which each participant likes the other(s). Positive feelings which are received from someone are reciprocated, i.e. the same degree of positive feeling is directed towards that person.

Recognition The second form of remembering identified by Ebbinghaus, and one which is used extensively by human beings. Ebbinghaus, working with lists of nonsense syllables, demonstrated that material which cannot be *recalled* may nonetheless be recognized as having been in a previously learned set of information, if it is presented to a subject. See also *reconstruction, relearning savings.*

Reconstruction Also sometimes known as redintegration, this is the third of the four basic forms by which memory may be demonstrated, according to the work of Ebbinghaus. Once subjects have learned a list of nonsense syllables, in the event of their being unable to *recognize* or *recall* the items learned, they are often able to reconstruct the list in its original sequence, if provided with the relevant items. Although they will not experience a specific memory of the list, one particular sequence often 'feels more right' than any other arrangement. See also *relearning savings, recall, recognition.*

Recreational drugs Drugs which are consumed primarily for enjoyment or appreciation of their effects, rather than for medicinal purposes. These include legal drugs such as *alcohol, nicotine* and *caffeine* and illegal drugs such as *marijuana, amphetamines,* and *heroin.* The use of recreational drugs in some form occurs in all known human societies, and in some cultures includes the use of very powerful *hallucinogens* such as *mescaline.* In general, the more powerful drugs are consumed within some kind of *ritualized* setting, while less potent ones such as marijuana are taken more casually. Within Western societies, however, the rituals are confined to sub-cultural habits, and are not often used as a framework for the experience of the drug itself.

Red–green colour blindness The most common form of colour blindness, in which the person is unable to distinguish between the wavelengths of red and the matching wavelengths of green. It has proved difficult to explain this in terms of the conventional *trichromatic theory* of colour vision, and the predominance of red–green colour blindness has been taken as important evidence for the idea of *opponent processing.*

Redintegration See *reconstruction.*

Reductionism A form of argument which takes the view that an event, behaviour or phenomenon can be understood as being nothing but its component or constituent parts. So, for instance, the insistence of the early *behaviourists* that human experience could be seen as nothing but combinations of stimulation–response links; or the view that behaviour may be understood as nothing but the action of 'selfish genes' are both reductionist arguments. Although often superficially appealing, reductionist argument ignores other levels of explanation, such as cognitive explanation or experiential/social factors in understanding the phenomenon, and as such can only provide a limited understanding of the event under study. Note that even if the most extreme reductionist position is true and all human functioning is the result of the activities of sub-atomic particles, it would be nonsense to try to explain a human activity such as a joke in these terms.

Redundancy A term used mostly in information theory for the extent to which a message does not provide new information. Redundant material, like the letters replaced by Xs in this senXencX, can XX put back quite easilX. Because language is highly redundant we can interpret messages accurately even when they are received in *noisy* conditions. In fact, the lower the *signal to noise ratio,* the more redundancy is needed in the message.

Reference group A social group which is taken by an individual as providing standards for the modelling of that person's own behaviour. The individual concerned may not actually belong to the reference group itself, but sees them as directly relevant to his or her own lifestyle or situation.

Referential Using words to refer to objects or events, but not with communication as a major objective.

Reflex A direct response to stimulation which occurs automatically, without any decision-making input from the *central nervous system.* For example the leg jerk which occurs when the knee is tapped. Reflexes are often referred to as involuntary responses, to distinguish them from the *voluntary* behaviour of deliberate action. They are usually mediated directly by the *spinal cord* rather than the brain itself, though this sub-group is sometimes identified as 'spinal reflexes'. See also *reflex arc.*

Reflex arc The term given to the sequence of neurones involved in the simplest unit of behaviour, the reflex. In its most basic form, the reflex arc consists of three neurones: the *sensory neurone*, which carries the information concerning the stimulus to the *spinal cord*; the *connector* neurone within the spinal cord which picks up the information from the sensory neurone and reroutes it; and the *motor* neurone, which passes the message from the connector neurone to the muscle fibres, causing them to contract and the reflex action to occur. Because reflex arcs follow well-defined paths, the failure to display an appropriate reflex can indicate precise forms of damage to the nervous system. Reflexes are therefore used to test for such damage, particularly in newborns where other responses are less available. See *Babinski reflex*.

Refractory period The period of a few milliseconds immediately after a neurone has fired and before it is completely restored to full functioning. The refractory period has two parts: the *absolute refractory period*, in which no amount of stimulation will make the neurone fire and the *relative refractory period*, in which the neurone will fire only in response to a particularly strong stimulus.

Regression Any return to a previous or simpler state. Particularly: (1) In *psychodynamic* theory, a retreat under stress to an earlier *psychosexual stage*; (2) A technique in therapy in which the patient is encouraged to think and feel as he did at a much younger age. The usual objective is to re-experience a traumatic event so that it can be properly dealt with in the supportive context of therapy. *Hypnosis* is often used to help the process; and (3) In statistics, measures of the extent to which one variable depends on another. Most commonly encountered in *linear regression*, the equation of the straight line which provides the best fit (or smallest total discrepancy) when *dependent variable* scores are plotted against the *independent variable*.

Regulator One of the types of *non-verbal* cue classified by Ekman and Friesen, regulators are those cues which regulate or structure social interaction. Examples of these are the time sequences and turn-taking of conversations, small noises such as 'uh-uh' made to indicate agreement during a conversation and to signify that someone is still listening and *eye contact*. See also *affect displays*.

Rehearsal A term used to mean practice, when applied to a memory task. Rehearsal is the repetition of the material to be learned.

Reification Treating ideas or concepts as if they were objects or facts, for example, starting from the fact that people can be seen to behave more or less intelligently, and going on to assume that there is a 'thing' called intelligence. It is easy to slip into reification when talking about cognitive processes: for example in Broadbent's *filter model* there is a box labelled 'filter', which is used to indicate a process. The mistake is to represent it as if it must be a mechanism. Another form of this error in psychology is to define a possible phenomenon and then assume it is a fact which then has to be explained. For example, there was a long period in which different theories were proposed to account for some children being obedient and others disobedient, before researchers observed real children and found that none were either consistently obedient or consistently disobedient. The fact that we have a good explanation for something (say, male aggressiveness) does not prove that the thing exists as an entity in itself, independently of the context in which it is manifest. See also *labelling*.

Reincarnation The belief that after death people are reborn either as another person or in some other animate form.

Reinforcement See *reinforcer*.

Reinforcement contingencies The circumstances under which reinforcement will be given. These may vary naturally or be systematically varied, as in the case of behaviour shaping.

Reinforcement schedule A particular pattern of applying *partial reinforcement.* Contin... There are four main types of reinforcement schedule, each of which produces a distinctive effect on the pattern of responding. Schedules may be either fixed or variable: if fixed, then reinforcement is given according to a predetermined pattern; if variable, it is given according to a randomized sequence which averages out at a particular number. Reinforcement may also depend on the number of responses that has been made since the last reinforcement, or the time interval which has elapsed since the last reinforcement was given. The four schedules are: *fixed ratio, fixed interval, variable ratio* and *variable interval.* Fixed-ratio reinforcement produces a rapid *rate of response* but a low *resistance to extinction.* Fixed-interval reinforcement produces a low rate of response and a low resistance to extinction. Variable ratio produces a high rate of response with a high resistance to extinction. Variable interval produces a steady, regular rate of response and a high resistance to extinction.

Reinforcer Something which strengthens a learned response; which makes a learned response more likely to occur again. In *classical conditioning*, the reinforcer is simply the repetition of the pairing of the unconditioned and conditioned stimuli. In *operant conditioning* the reinforcer is the event that occurs after the operant behaviour, making it more likely to occur again, and which may be either *positive* or *negative.*

Related-measures design A subject design used in experiments in which the same subjects are used in both the experimental and the control conditions. Since each subject's score is compared with one obtained from the same subject, this technique allows the experimenter to control for individual subject differences, e.g. in IQ level or motivation. However it does mean that *order effects* are likely to become important in the study, and so related-measures designs often involve the use of *counterbalancing* as a control. It is also known as a repeated-measures design, or a correlated-subjects design. The paired *t-test* is used in related-measures designs only.

Relative refractory period The period after a neurone has fired when it will only respond to a stimulus of unusual strength. This occurs after the *absolute refractory period*, when it will not fire at all, and reflects the cell's renewal of resources after the production of the burst of electrical energy in the form of the *electrical impulse.*

Relative threshold The degree by which a stimulus must increase in order for the increase to be perceived. The *threshold* is set at the point where 50% of changes of that magnitude are perceived, and changes in direct proportion to the intensity of the initial stimulus. The law known as *Fechner's law* expresses this relationship. See *just noticeable difference.*

Relaxation training A range of techniques to bring about a relaxed state in the subject. Usually used as a component in therapy, for example in maintaining a relaxed state in a *phobic* patient as they approach the feared object. Many of the techniques used in psychotherapy are based on methods developed for meditation, such as yoga, or are variations on hypnotic induction procedures. Edmund Jacobson popularized the approach with a procedure in which the subject concentrates on, and relaxes, groups of muscles in turn. *Biofeedback* can also be used.

Relay neurone A neurone found within the *spinal cord* and the *brain*, which forms multiple connections with several other neurones, and allows for information to be routed in several different directions simultaneously. Relay neurones are also known as *connector neurones* or *multipolar neurones*.

Relearning savings The fourth (weakest) level of remembering identified by Ebbinghaus in his work on the memorization processes. He found that there were situations where all traces of memory of a specific set of items appeared to have been lost, in that the set could not be *recalled*, recognized, or *reconstructed*; but when the set of items was encountered again, it would take less time to relearn than a comparable set which had not previously been learned.

Releaser Something which acts as a signal to trigger off a particular response – usually an inherited one. See also *sign stimulus, IRM*.

Reliability The consistency of a measure, how likely it is to produce the same results if used again in the same circumstances. Reliability is a significant concern in the development of *psychometric tests*, and is usually assessed by one of three methods: test–retest, in which the same test is administered to the same subjects after a period of time has elapsed; split-half testing, in which the score that the person achieves on one half of the test items is compared with that obtained on the other half (both administered at the same time), to see if they give similar outcomes; and alternate-forms testing, in which two *matched* versions of the test are given to the same subjects on two different occasions, with their results being compared. Reliability is an essential requirement of any measure, whether a test, a physiological measurement, an observational procedure or whatever. The other essential requirement is *validity*. A major difficulty in assessing reliability is that scores on successive occasions may differ either because of *practice effects* or because the subjects have actually developed or changed in some other way. George Kelly said that reliability is a measure of how insensitive a test is to people changing. More broadly, reliability should be demonstrated over the period for which the function being measured is believed to be stable.

REM sleep See *rapid eye movement sleep*.

Repair process An aspect of speech in which attempts are made to correct a misleading utterance or some other source of misunderstanding.

Repeated-measures design See *related-measures design*.

Repertory grid A technique developed by George Kelly, for utilizing a person's *personal constructs* to examine the significant people in his world, and so identify actual or potential sources of psychological discomfort or stress. The repertory grid is an *idiographic* technique, which enables a therapist to see the patient's world as they see it, a valuable first step in most forms of therapy. The repertory grid is also used more generally in research to indicate how people perceive and understand their worlds.

Replication Repeating an experiment to ensure that the results are reliable, and not due to the particular circumstances or chance at the time of the first experiment. Psychology experiments are particularly open to influence from incidental factors, like the expectation of subjects at a particular time and place, and so should always be replicated. However they are not, for a variety of reasons; it is difficult to get grants for replications, difficult to get them published, and most experimenters would rather run their own, new, experiment than someone else's old one. The result is that many of the most famous findings have never been replicated and are not reliable.

Representative sample A sample of subjects in a study which has all the important characteristics of its parent *population*, so that it can be regarded as typical of that population for research purposes. There are several different techniques for obtaining a representative sample which include *quota sampling* and *random sampling*.

Repression A *defence mechanism* by which unacceptable thoughts or desires are forced into the *unconscious*. As with all defence mechanisms, the psychological relief is paid for by having to distort one's perception of reality.

Resistance In psychotherapy, the attempts by the patient to prevent the therapist being effective. In analysis particularly, resistance is seen as an inevitable response of the unconscious to the therapeutic process of making it conscious. When a patient rejects one of the therapist's *interpretations*, this will be regarded as resistance and may be taken as an indication that the interpretation was approaching a particularly important *defence*. However it must sometimes be the case that patients reject interpretations simply because they are wrong.

Resistance to extinction How long a learned response will carry on without any further *reinforcement*. Resistance to extinction is often used as a measure of *operant strength*, in other words, to indicate how strongly something has been learned.

Response bias The tendency that subjects have to produce experimental responses which are socially desirable, or that they think the experimenter expects. For example, a study involving comparing reactions to sexually explicit material with reactions to neutral material may show a difference which results from the subject's unwillingness to appear overly concerned with sexual matters; or from their embarrassment. If this is not directly the topic under study, it will result in a response bias which could obscure other experimental findings. See also *experimenter effects, confounding variables.*

Response generalization The tendency to produce a learned response in conditions which are similar, though not identical to those under which the response was learned. In general, the more similar the conditions are, the stronger the response will be, known as the *generalization gradient.*

Response rate How frequently a response or unit of behaviour occurs in a set period of time. Response rate is often used as a measure of *operant strength*, or as an indicator of how strongly something has been learned.

Resting potential The chemical balance between the external fluid and the chemical components of a neurone when it is not firing. See also *action potential.*

Restricted code A code of language use identified by Bernstein, which is characterized by a high proportion of personal pronouns, a relatively limited vocabulary, and a considerable reliance on shared assumptions on the part of the speaker and listener. Bernstein saw restricted-code speech as mainly used by working-class individuals, whereas its counterpart, *elaborated code*, was mainly used by middle-class people. Because of the high dependency on context in restricted-code speech, Bernstein argued that this made its speakers less likely to deal with abstract concepts and related forms of knowledge: a version of the *verbal deprivation hypothesis* which was highly criticized, notably by Labov, who showed that users of restricted codes demonstrated abstract reasoning just as readily as elaborated-code users.

Reticular activating system (RAS) A region of the lower portion of the brain which

Direction of light

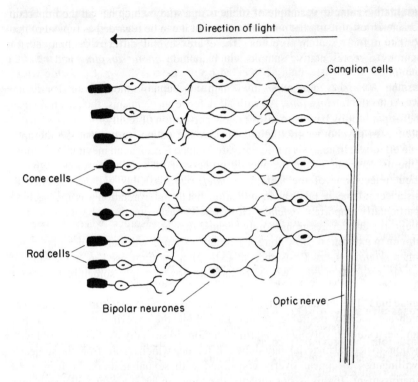

Ganglion cells

Cone cells

Rod cells

Bipolar neurones

Optic nerve

The structure of the retina

is directly involved in attention, sleep, and wakefulness. The RAS appears to operate as a kind of switching mechanism for whole areas of the *cerebral cortex* in the event of wakefulness and alertness; and its surgical removal results in permanent unconsciousness.

Reticular formation See *reticular activating system*.

Retina The three cell deep layer on the back inner surface of the eyeball. The layer furthest away from the lens consists of light-sensitive *rod cells* and *cone cells*, the next layer consists of *bipolar neurones* and the third layer consists of ganglion cells with elongated *axons* which cross the retina and join together at the blind spot to form the *optic nerve*. The retina forms a 'screen' on which an image is projected from the pupil of the eye; the light information which falls on the retina is converted into electrical impulses by the rod and cone cells. The point on the retina is converted into electrical impulses by the rod and cone cells. The point on the retina where the image is focused most sharply is known as the *fovea*, and corresponds to the point where visual attention is concentrated in normal perception; the rest of the retina provides the remainder of the *visual field*. [f.]

Retinal image The camera-like image of the external world which is cast on the retina by light rays entering the eye through the *pupil*, and focused onto the retina of the eye by the lens. The retinal image is inverted, but is a projected picture of the external world which falls on the layer of *rod* and *cone cells* in the retina.

Retinal size The term used to refer to the image which an object casts on the retina. This will vary in proportion to the distance away of the object: a six-foot pole viewed from 40 yards will cast an image whose retinal size is half that of a six-foot pole seen from 20 yards. However, due to the process of *size constancy*, the size of the object which is actually perceived by the person does not coincide with its retinal size, and even in young infants some amount of compensation for distance seems to occur.

Retrieval A term used to refer to the process of remembering things, in which the information is seen as being 'retrieved', or brought back from, some kind of storage system.

Retroactive interference When new information which is being learned interferes with the ability to recall information which was learned previously. For example, a tennis player who takes up squash may find their tennis deteriorates for a while. See also *proactive interference, transfer of training*.

Retrograde amnesia The form of *amnesia* (memory disorder) where the person is unable to remember things which happened before the event which rendered them amnesiac. Retrograde amnesia usually occurs after some form of brain damage, but can happen in a minor way after concussion. It is not uncommon for people who have been in an accident involving severe concussion to lose all memory of the few minutes leading up to the accident. See also *anterograde* amnesia.

Retrospective study A study which involves collecting data about events which happened in the past. Many *epidemiological* studies have been retrospective, using data from the previous records of patients or clients. The weaknesses of this technique, are the inadequacy of documented information in the recording of salient or influential events in a person's life and the tendency for researchers to focus exclusively on the particular feature that they are interested in, and to ignore other information. Where straightforward research on medical conditions is concerned, this may not be a problem; but the technique has been used for far broader research, notably Bowlby's work on *maternal deprivation*, which greatly influenced views of child-care practice from the 1950s to the 1970s.

Reversibility The *operation* of returning something to its original state by reversing the process which transformed it in the first case. The concept of reversibility plays an important part in Piaget's theory of cognitive development. Understanding that an operation is reversible allows one to understand important aspects of the world. Examples are: if a ball of plasticine can be rolled out into a sausage shape, it can also be rolled back into a ball. If A is larger than B, then B is smaller than A. If 3 squared is 9, then the square root of 9 is 3. Piaget saw an understanding of reversibility as an essential part of *concrete operations*: in particular it is necessary before *conservation* can be acquired.

Reward Something which is provided for an organism, animal or human, after a desired piece of behaviour has occurred and which takes the form of something that the organism wants, needs, or likes. The concept is particularly important in the theory of *operant conditioning*, where reward forms *positive reinforcement* for learned behaviour.

Rhetoric The study of how language is used to persuade others. Rhetoric was an important part of education from the early Greeks to the mid 19th century, but came to be seen as inferior to the search for scientific certainties. As it explicitly recognises the extent to which thinking is affected by ideology, it is favoured by certain *social constructionists*.

Rhine cards See *Zener cards*.

Ribonucleic acid (RNA) A chemical found in the cells of the body, which is involved in genetic protein synthesis, and is capable of duplicating genetic material, *DNA*, for use elsewhere. It has also been thought to be involved in learning processes.

Right hemisphere The half of the *cerebrum* to the right side of the head. It is mainly concerned with the functioning of the left side of the body, and the right side of the retina in each eye. Following a series of *split-brain studies* by Sperry, it was found that this half of the brain was particularly adept at spatial and artistic tasks, whereas the *left hemisphere* was more readily concerned with language and number.

Risky shift The finding that when a group of people makes a decision it tends to be riskier than the decision that they would each have made individually (riskier than the average of the individual decisions). There are several possible explanations for the risky shift, one being that it is an example of *diffusion of responsibility*; and a second being the 'risk as value hypothesis', that risk taking is socially valued and so people will want to be seen by the group as more risky. However, some psychologists question whether it really happens. See also *group polarization*.

Rite of passage A *ritual* which marks the progress from one stage of life to the next. All societies have their own rites of passage, with weddings, funerals, and those marking the transition from childhood or adolescence into adulthood having been most studied by anthropologists.

Ritual A strictly defined pattern of behaviour which carries a significant social meaning in well defined context. Marriage ceremonies are a clear example of culturally defined rituals but the term is used more widely to include any meaningful patterns of behaviour carried out according to strict rules such as the hand washing ritual of an *obsessional* person, Sunday dinner, or a task that a family might be asked to undertake as part of therapy.

RNA See *ribonucleic acid*.

Robotics The area of research which involves the development of mechanical systems which can perform a set of actions in a way comparable to that of a human being. Many highly successful robotic systems have been developed and applied, particularly in the manufacturing industries. They have involved considerable research not just into movement systems but also into the development of such techniques as optical scanning devices, which can identify and respond to anomalies or changes in the appearance of the material being manufactured. As such, robotics is often considered to form one branch of the research into *artificial intelligence*.

ROC curve See *receiver-operating-characteristic curve*.

Rod cells Light-sensitive cells in the *retina* of the eye, which respond to very small amounts of light, but are not sensitive to colour. Rod cells are found in all parts of the retina except the *fovea*. Their action is most apparent at the edge of the retina, where their extreme sensitivity provides acute detection of movement in the peripheral vision, and allows very faint objects to be seen. Night vision is due to the sensitivity of rod cells.

Rogerian A term applied to methods of counselling or psychotherapy which are based on the work of Carl Rogers. See *non-directive therapy*.

Role The part that each individual is expected to play in a social situation. This has been studied particularly in groups in which a role is likely to be allocated to each member: leader, fixer, clown, loyal member, etc. Any individual is likely to play different roles in different groups and may therefore experience role conflict when two groups come into contact, for example when adolescents encounter their family while in the company of their gang. Roles may be held very briefly (e.g. the one who has the next turn), over long periods (child), or permanently (gender role). See the next five entries.

Role behaviour Behaviour which is considered to be appropriate for someone who is playing a specific social role. For instance, someone playing the role of a shop assistant is expected to behave in certain ways, to be smart, and alert, and to demonstrate specific behaviours such as asking if a customer needs to be served or requires information about prices, etc. Other kinds of behaviour of which the person may be equally capable, such as ballroom dancing, are completely inappropriate to the social role of shop assistant. See also *role expectation*.

Role confusion In Erikson's developmental theory, a state in which the *identity* is not well defined. It may be regarded as a temporary state (this can occur at any time of life but is particularly common during adolescence) or as the long-term consequence of having failed to establish a clear identity during adolescence. See *psychosocial stages*.

Role count The sum total of social roles which an individual plays. The concept becomes particularly important in the case of those who have recently retired; the process of retirement results in a drastic reduction in the number of social roles played by the individual, and some researchers consider that it is important for the retired person to replace at least some of those social roles in alternative social activities. See *disengagement*.

Role expectation The implicit but nonetheless very clear ideas which members of a society have concerning the ways that people ought to behave when they are playing a social role in that society. Behaviour which does not conform, at least in general terms, to role expectations will usually meet with social sanctions of some kind, for example the exclusion of the person from the group.

Role play Taking a particular role temporarily and behaving, as nearly as possible, like a person who actually holds that role. Role play is widely used in training situations and is an effective way of helping people understand what it feels like to have the given role, and allows them to practice the role before being fully committed to it. It has been found that acting a role often shifts a person's opinions towards those they have been working with. Preparatory role play may also help reduce anxiety and improve performance in stressful situations such as interviews.

Rorschach ink-blot test A *projective test* based on psychoanalytic theory, in which subjects are shown large and elaborate ink-blot patterns, and invited to interpret them in terms of images which the blots might represent. The idea is that the responses which subjects make will indicate the concerns of the *unconscious* mind. The Rorschach has been found to have poor *reliability*.

Rosenthal effect The finding by Robert Rosenthal and others that one's expectations can have an effect on an outcome that is being observed. Used particularly in connection with the finding that when teachers were told that a group of children were very bright, those children performed better than a similar group that the teachers had been told were all dull. The term is also used for various forms of *experimenter effect* and *self-fulfilling prophecy*.

S

Saccade Rapid, unconscious jerks and tremors which are made continuously by the eye and are thought to be instrumental in preventing *habituation* of the *retinal image*.

Sadism A psychosexual disorder in which a person obtains sexual arousal by inflicting pain or humiliation on another person. See *masochism*.

Safety need The second level of Maslow's hierarchy of needs, safety needs refer to needs for security, shelter and freedom from attack. These needs become important once basic *physiological needs* have been satisfied. Once the safety needs in turn have been satisfied, according to Maslow, the next level of needs, *social needs*, become important.

Salience Something which is particularly noticeable or likely to be perceived. The salience of an object or event may be due to its physical properties, such as brightness and clarity; or it might arise because the object or event relates to needs, emotional states or meanings on the part of the perceiver.

Sample A part of a *population*, which is studied so that the researcher can make generalizations about the whole of the original population. Samples can be gathered by means of several different procedures, which include *quota sampling* and *random sampling*. Nearly all psychological research is carried out on samples, because the size of populations, or some other factor, makes studying the whole impossible. Many statistical techniques are concerned with indicating the reliability of a conclusion based on a sample, but cannot identify whether the sample is typical of that population or not. So a considerable amount of experimental methodology is concerned with ensuring, as far as possible, that the samples involved in the study are representative.

Sampling procedure The procedure by which a sample is acquired. Sampling procedures need to be carefully defined and reported so that it is possible to judge whether the results from that sample can be generalized to the *population* or to other samples. The commonest form is *random sampling* in which members of a population are selected at random, with each having had an equal chance of being chosen. In practice, truly random sampling is difficult to achieve because of such influences as volunteer bias. A further disadvantage is that a random sample needs to be quite large to ensure a close fit to the parent population. More sophisticated forms such as *stratified sampling* are designed to represent the population on all important aspects and so allow reliable conclusions to be drawn from a smaller sample.

Sanction Some kind of negative event which occurs as a result of undesired behaviour. Sanctions may include the withdrawal of privilege or opportunity as well as direct punishment or other unpleasant consequences.

Satiation Satiation is defined operationally as the point at which an animal will no longer seek food. It is usually used in investigations of hunger or other motivational states and it implies that the underlying *need* is temporarily satisfied.

Saturation This term is usually used with reference to colour, and indicates how 'rich' the particular hue concerned is. But it may also refer to other forms of stimuli, and in general it is concerned with intensity of content. So, for instance, 'saturation advertising' is when an advertising campaign is so intense that it is considered to have achieved the maximum possible exposure to its target audience.

Scaling The process of organizing recorded measures into a scale. By doing this, the measures can be given values with known arithmetical relationships to each other, and statistical analyses can be undertaken. Scaling is particularly important in psychology because many of our phenomena cannot be measured directly, being either subjective or too complex. See *levels of measurement*.

Scapegoat theory The idea that prejudice arises from people seeking to blame others for their own negative circumstances. According to scapegoat theory, poor living conditions, economic depression and frustrating situations lead people to react in hostile ways to others; and this reaction is likely to focus on any individuals who are present but don't belong to the person's own *peer group*. Scapegoat theory has been put forward as an explanation for the growth of *racism* and *sexism* during times when economic circumstances are difficult.

Scattergram A diagram used to illustrate *correlations*, in which the vertical axis (the *ordinate*) represents the values of one set of scores, and the horizontal axis (the *abscissa*) represents the other set. Each pair of scores is plotted as a point on the diagram. This means that the relationship between the two variables can be seen in the way that the scores are scattered within the area described by the diagram. [f.]

Schedule of reinforcement See *reinforcement schedule*.

Schema A hypothetical model of the way that information is stored by the brain. It is used to direct action, and in understanding the relationships between events. A schema would include all of the information relating to a particular event or type of event, including representations of previous actions; theoretical and practical knowledge about the event; ideas and opinions about it, etc. The concept of schemata has formed a major part of the theory of *cognitive development* put forward by Piaget, and the cognitive theory of Neisser. The idea of schemata, and their extension and development through experience, provides a useful model for understanding how many different levels of comprehension can be involved in both new and familiar situations. See also *anticipatory schema*.

Schizoid Showing tendencies towards *schizophrenia*, but not the extreme form. The term is therefore used for any indication of a mismatch between thought and feeling; and a lack of interest in, and ability to form, social relationships.

Schizophrenia A broad group of *psychoses* in which emotion is blunted or is not coordinated with thought and behaviour, and in which thought appears to be disordered. Typical symptoms of schizophrenia are *hallucinations*, incoherent speech and thought, and *delusions*. While suffering from schizophrenia a person is unlikely to maintain social relationships or to look after themselves adequately. There is considerable controversy over whether schizophrenia is an *organic* disorder, or whether it happens as a result of life experiences, particularly growing up in certain kinds of family. One of the reasons why this dispute has not been resolved is that there is no very clear and widely accepted definition of schizophrenia, and diagnosis of the condition varies widely in different countries.

School refusal (school phobia) The insistence by a child on staying at home rather than going to school. It used to be thought of as resulting from a fear of the school, hence 'school phobia', but is now often seen as being a fear of being away from the home. This is usually to do with either the fear of some disaster happening to the mother or of the parents leaving during the child's absence. Sometimes these fears have a rational basis: for example, when there is serious discord between the

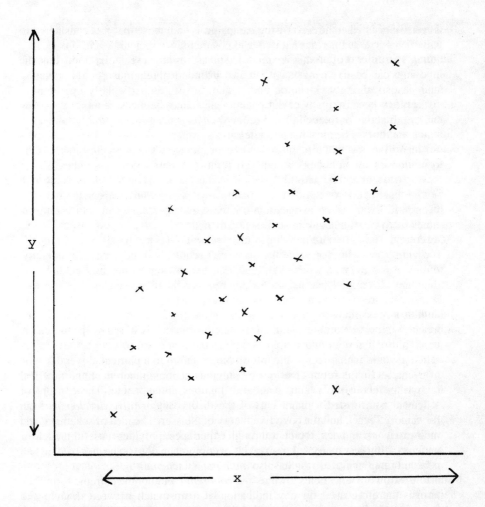

A scattergram showing a positive correlation

parents or when the child has often been threatened with abandonment. See also
separation anxiety, insecure attachment.

Script A concept which was particularly developed by Schank and Abelson (1977),
scripts refer to the implicit set of social expectations and assumptions which operate
in the course of everyday interaction. People act in accordance with these
expectations, as if their part was 'scripted'. The classic example of this is the
restaurant 'script', in which the roles of customer, waiter, etc. are highly prescribed,
with certain actions being expected at certain times. Schank and Abelson
demonstrated that the application of a particular script to a given situation can
channel and structure what is perceived from that situation: a different script will
lead to the individual noticing different features and remembering different facts as
salient. Scripts are for the most part unconscious and assumed: it is noticeable that
people are more likely to remember deviations from a script (e.g. not being asked

if coffee is wanted at the end of a restaurant meal) than they are to remember events which conform to the script itself. See also *schema, social representations*.

Secondary circular reactions *Circular reactions* which have progressed beyond involving the infant's own body, and now operate through manipulating objects, again in a fixed repeated pattern.

Secondary process In psychoanalytic theory, secondary process refers to conscious rational thought. It is called this in order to distinguish it from *primary processes* which are the workings of the *unconscious*.

Secondary reinforcement Something which has acquired the property of being able to *reinforce* learned behaviour, because it has previously been associated with a *primary reinforcer*. As an example: if a 'click' is sounded each time a rat in a *Skinner box* is rewarded with a food pellet, the noise becomes associated with the reward. If the behaviour is then allowed to *extinguish*, it can return and be maintained simply by the giving the sound of the 'click', with no food reward being necessary. Also, a new behaviour can be increased if it is just followed by the click. The noise has developed reinforcing properties, and become a secondary reinforcer, simply as a result of its association with the primary reinforcer. There are many different kinds of secondary reinforcer: in human terms, the most frequently encountered one is almost certainly money.

Secondary sexual characteristics Physical characteristics such as beards and breasts, which are normally found in mature members of one sex only, but which are not the actual sex organs. See also *primary sexual characteristics*.

Sedative A drug which has a calming effect on the individual, usually producing drowsiness. This is often achieved by dampening the *autonomic* activity of the *nervous system*. Sedatives are known to produce considerable *tolerance* in the body, with progressively increasing amounts of the drug being required to produce the same effect. The most well-known sedatives are the *barbiturates*, which traditionally have been prescribed as sleeping tablets, though this is now less common. Although there are known to be large numbers of people who are addicted to barbiturates, the addiction has usually been induced through medical prescription: they are not commonly used as *recreational drugs*.

Selective attention Attention which is channelled towards certain stimuli and ignores the presence of others. The most well-known example of this is when someone is concentrating on one particular conversation among a large amount of background noise, some of which may be actually louder than the conversation being attended to. This was dubbed the *cocktail party effect* in the 1950s, and gave rise to a considerable amount of research, often involving *dichotic listening tasks* and *split-span tasks*. The research gave rise to several different *filter theories*, which eventually showed that there is a considerable amount of unconscious *semantic* processing even of unattended information.

Self-actualization A concept central to the *humanistic* theories of both Maslow and Rogers although used in a different way by each. Broadly speaking, self-actualization refers to the making real (actualizing) of human potential. So it involves the individual developing their abilities to the full, exploring options and skills, and experiencing life as fully as possible. For Maslow, self-actualization takes the form of a 'peak experience', which is only attained once all of the 'lower' levels of the hierarchy of needs have been satisfied, needs such as *safety needs, physiological needs*, etc. Accordingly, self-actualization is seen as a relatively uncommon event, which occurs only in a few special individuals.

In Roger's theory, by contrast, self-actualization is seen as a continuous process of self-exploration and development which forms an undeniable need for the individual. Most people have ways of developing their potential in day-to-day living, through hobbies, interests and the like, and most recreational pursuits involve some degree of trying to learn or to improve one's abilities. In some individuals, though, the need for self-actualization comes into conflict with the need for *positive regard* from others. Self-exploration is seen as potentially threatening, in that it might incure disapproval and censure from other people. Accordingly, such people suppress their need for self-actualizaton, and Rogers sees this as forming the foundation of *neurosis*, because the person experiences a discrepancy between the way that they actually act, and their 'inner self'. But if they have a relationship involving *unconditional positive regard* from someone, the person becomes able to explore their need for self-actualization, and to balance the two needs in such a way as to achieve personal growth and maturity. Providing such unconditional positive regard forms the basis of Roger's *client-centred therapy*.

Self-concept The sum total of the ways in which the individual sees her or himself. Self-concept is often considered to have two major dimensions: a descriptive component, known as the *self-image*, and an evaluative component, known as *self-esteem*, although in practice the term is more commonly used to refer to the evaluative side of self-perception.

Self-consciousness An exaggerated awareness of one's own behaviour, feelings and appearance, combined with a belief that other people are equally aware, interested, and critical. Self-consciousness is often particularly extreme during adolescence.

Self-efficacy beliefs The belief in one's own power to act effectively, or to influence events. Particularly associated with the work of Albert Bandura (e.g. Bandura, 1977), self-efficacy theory argues that high self-efficacy beliefs contribute directly to a positive sense of agency in dealing with the world. It is therefore closely linked with an internal *locus of control*. People with high self-efficacy beliefs have been shown to make more efforts to achieve results, and to respond productively to *feedback*; those with low self-efficacy beliefs show a tendency to give up easily, and to fail to use feedback to improve their performance. Although they are closely linked with, and perceived as a major contributor to, *self-esteem*, self-efficacy beliefs can be highly specific, relating only to particular types of task. There is, however, some suggestion that people do show a general tendency towards high or low self-efficacy beliefs in a wide range of contexts. Bandura argues that it is often psychologically healthier for an individual to have slightly higher self-efficacy beliefs than the evidence would warrant, since that will encourage them to take on more difficult tasks, and to persist at those tasks in the face of initial difficulty. This in turn increases their likelihood of success. Some developmental psychologists believe that a strong sense of self-efficacy is built up in infants and small children through *contingencies* provided by caregivers.

Self-esteem The personal evaluation which an individual makes of her or himself; their sense of their own worth, or capabilities. Excessively low self-esteem is regarded as indicating a likelihood of psychological disturbance, and is particularly characteristic of *depression*. There are several simple questionnaires which have been developed for measuring self-esteem, as well as more sophisticated tests such as the *Q-sort*.

Self-fulfilling prophecy A statement which comes true as a result of having been made. The classic example of the self-fulfilling prophecy in action came from work by Rosenthal, in which undergraduate students were given a set of experimental rats to train in maze-running. Despite the fact that there were no observable behavioural differences between the rats at the start of the experiment, the students were told that they could expect some to be very quick at learning the maze, while others would be very slow. The rats performed according to these predications, because the predications had induced expectations on the part of the students which affected how they handled the animals during training. Further studies by Rosenthal and his colleages demonstrated the power of expectations held by teachers towards their pupils, and the self-fulfilling prophecy is now considered to be a major social influence which needs control in psychological investigation. See also *experimenter effects, double-blind control.*

Self-image The internal picture which an individual has of her or himself; a kind of internal description, which is built up through interaction with the environment and feedback from other people. The self-image may include knowledge about hair colour, (though not attitudes towards it), and the *social roles* played by the individual. The person's attitude to self-image plays an important part in their level of *self-esteem.* Most people operate a self-image which gives an exaggerated idea of their own attractiveness and this seems to be necessary for psychological well-being. See also *body image, identity.*

Self-perception theory The idea that we gain knowledge about ourselves by observing our own behaviour, e.g. 'I must have been hungry because I ate an extra sandwich'. Overtly such an approach may appear naive, yet there is considerable evidence to suggest that people do make *attributions* about their own behaviour based on how they have seen themselves acting or reacting.

Self-persuasion The modification of a person's beliefs to become consistent with what they observe about their own behaviour.

Self-serving bias A bias in a person's thinking which serves one of their purposes such as maintaining *self-esteem* or cognitive consistency. The concept is used particularly in *attribution* theory, to refer to causal beliefs that are adopted because they are favourable to the individual.

Semantic To do with meaning; the intended communication or meaning which underlies any utterance or signal. The word semantic is usually used in contrast with *syntactic,* referring to the structure of the communication (e.g. sentence structure). Such contrasts are particularly useful in examining the use of *language* in communication.

Semantic conditioning A conditioning process which uses a *stimulus–response* form of learning like *operant* or *classical conditioning,* in which the individual is trained to respond to the meaning of a word or phrase. Although the perception of meaning is a *cognitive* rather than a *behavioural* event, studies of semantic conditioning are reported to show all the characteristics of behavioural conditioning, such as *generalization, discrimination,* etc. However, there is a certain amount of evidence to indicate that semantic conditioning only 'works' if the subjects catch on to what the study is about, and decide to cooperate.

Semantic memory Memory that is concerned with processes: how to do things. For example, people with amnesia may forget knowledge-based information, but they rarely forget such things as how to walk, boil a kettle, or write. These are examples of semantic memory. See also *episodic memory.*

Semantic relations grammar A theoretical approach to understanding the way in which very small children put words together, which emphasizes the meaning, or intention, underlying the utterance. The short sentences and limited *utterances* of the child are viewed as *telegraphic speech*, signalling the most important parts of the communication, and only becoming more refined in terms of additional words or word endings later on. The theory was developed by Roger Brown in opposition to the view of language acquisition developed by Chomsky, which largely ignored what the child was intending to communicate and concentrated instead on the structure of the utterance. See also *psycholinguistics*.

Semicircular canals Structures in the *inner ear* which detect the overall movement of the body, and are particularly concerned with the sense of balance. The canals are filled with a fluid which contains in suspension small bony particles known as otoliths. As the fluid moves in the canals, the otoliths make contact with hair cells which line the edges, which then produce an *electrical impulse*. This is then passed to the brain, particularly the regions of the *cerebellum* which are concerned with balance and equilibrium.

Semiotics The study of patterns in communication of all kinds, including *language, ritual, non-verbal communication,* animal communication, etc. Although primarily concerned with the meanings within such communication, the study of semiotics also sees the form of the communication as providing important clues to that meaning. In other words, a clear distinction between meaning and form is not considered appropriate, as the form will influence the meaning, and the intended meaning will affect the choice of the form. For example, a reminder to staff in an office from the boss about switching off unnecessary lights could be delivered as a spoken communication, a hand-written memo, or a formally typed memo. Although the words might be identical, the form affects the meaning of the communication.

Senile dementia A loss of intellectual capacity which occurs apparently through a deterioration of the brain. The deterioration may not be directly attributable to ageing, and is often due to degenerative conditions such as *Alzheimer's Syndrome*, which can also affect younger people.

Sensation Anything which is experienced through the senses: a general term which is used to refer to sound, visual experiences, smell, taste, tactile or kinaesthetic experiences. It is usually used when it would be inappropriate or misleading to describe the particular form that the experience will take or has taken.

Sensitive period A time period during development in which a given capacity or form of learning can be acquired more easily. Sensitive periods are distinguished from *critical periods* by the fact that the capacity can be acquired outside the set period, though with greater effort.

Sensori-motor stage The first of Piaget's four stages of *cognitive development*, in which the immediate cognitive task which the child faces concerns the decoding of sensory information, and the coordination of motor action. The first step in achieving this, according to Piaget, is the reduction of the infant's *egocentricity* to the point where it can distinguish betwen 'me' and 'not-me', and has formed its first *schema,* the *body-schema*. Another important milestone during this period is the development of *object constancy*. See also *pre-operational stage, concrete operational stage, formal operational stage*.

Sensory adaptation The process by which our senses adjust their sensitivity to the surrounding environment. For instance, at night when background sound levels

Sensory neurone

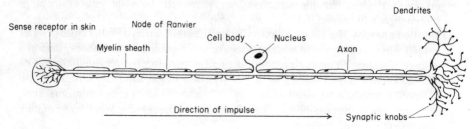

A sensory neurone

tend to be low, the ear will detect sounds which are far fainter than those which can be detected during the daytime. Similarly, the *rod* and *cone cells* of the *retina* become more sensitive in dim light, and less sensitive in bright light. See also *dark adaptation*.

Sensory neurone A neurone which carries information from a sense organ or sensory receptor to the *central nervous system*. Sensory neurones are usually bipolar, which means that the cell body occurs in between the two ends, each of which branches into *dendrites*. They are also *myelinated*, which allows them to transmit information extremely quickly. [f.]

Sensory projection area Areas of the *cerebral cortex* which receive sensory information, usually via the *thalamus*. There are four major sensory projection areas on each *cerebral hemisphere:* the *somatosensory area,* the *visual cortex* (also sometimes referred to as the striate cortex), the *auditory cortex*, and the *olfactory cortex*. As there seems to be some kind of correlation between the amount of stimulation and the amount or region of the sensory areas stimulated, it was

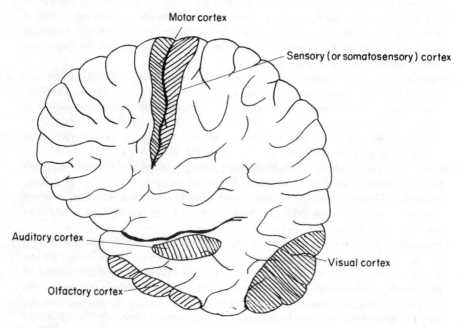

Areas of the cerebral cortex

originally thought that the sensation was 'projected' onto the area as if onto a screen, hence the name. [f.]

Separation anxiety The signs of anxiety and distress shown by a young child or other animal when a caregiver to whom they have an *attachment* leaves them alone in a strange situation. Traumatic experiences of separation, or repeated threats of abandonment ('if you're bad we'll put you in a home') in early childhood are believed to produce 'separation anxiety' in which the child is in a continual state of anxiety about the possibility of losing its primary caregivers. *School refusal* may also reflect separation anxiety.

Serial Occurring in sequence, one item at a time. So, for instance, a serial memory search is when the memory is searched for items in accordance with a definite pattern, one item after another.

Serial position curve A graph which shows the probability of recalling an item against the position that item had in the original list of items that was learned. The curve shows higher probabilities for the earliest and latest items in the list, indicating the effects of *primacy* and *recency*. The curve has also been used to distinguish the operation of *long-term* and *short-term memory*, although this is rather more contentious.

Serial processing The processing of information one item at a time. Many early *cognitive* models assume serial processing in, for instance, *problem-solving* or the decoding of *language*, although recent evidence suggests that, in fact, information is often processed on several levels simultaneously (*parallel processing*).

Serial reproduction A technique for investigating *constructive memory* developed by Barlett, in which a first account is reproduced from memory, and then the remembered account is again reproduced from memory, and so on. In this way, errors and alterations which occur in the accounts become cumulative, and therefore easier to classify and categorize. One everyday example of the use of serial reproduction is in the game 'Chinese whispers', in which a sentence or phrase is passed along a line of people, each person passing the message on by whispering it into the ear of the next person. By the time the message reaches the end of the line, it has usually become completely distorted.

Serotonin A *neurotransmitter* which is involved in a variety of brain processes, especially those concerned with moods, and including *motivation*, sleeping, relaxation, and pain. The *hallucinogens LSD* and psilocybin appear to be picked up at serotonin receptor sites, although the precise mechanisms by which these drugs operate is not fully understood. In some texts, serotonin is referred to as 5-hydroxytryptamine, or 5-HT.

Set A state of preparedness or readiness for a particular type of experience. Set may be demonstrated with most forms of *cognitive* process, but the most striking examples of it are *perceptual set* and *learning set*. In each case, information which is relevant to the prepared state is picked up far more quickly and easily than information which is not relevant.

Set-weight A pre-determined body weight, which seems to form the 'natural' weight of the animal concerned. The idea of set-weight arose from studies of the *hypothalamus*, in which it was observed that rats with *lesions* in particular areas of the hypothalamus would eat more than usual. At first it was thought that these areas represented 'feeding centres', but later findings showed that the increased intake only lasted until they had reached a certain body weight. At that point the rats would adjust their food intake to stay at that level. Experimental lesions in other areas of the hypothalamus produced effects in the opposite direction: rats would cease to eat until their body weight had dropped to a certain point, whereupon they would resume eating but eat only enough to maintain the new body weight. It has been suggested that similar mechanisms might be implicated in the case of obesity in humans.

Sex differences Differences between the sexes. Some psychologists reserve this term for biologically based differences, with *gender differences* being used for socially derived features. However the distinction is difficult to apply in practice and it seems likely that few differences between the sexes are either purely biological or purely social in origin.

Sex-linked trait An inherited tendency which appears only in members of one sex. The genes for such traits are carried on the pair of chromosomes which determine the biological sex of the individual. Because the structure of this pair of chromosomes differs substantially in males and females, sex-linked traits operate differently for the two sexes. For example, colour blindness which is *recessive* and carried on the *X chromosome* only will only be apparent in females who have inherited it on both of their X chromosomes. If it is carried on only one of them, then the gene for normal colour vision on the other X chromosome will dominate. However, colour blindness will always appear in males who carry it on the only X chromosome they have. There are therefore more colour-blind men than colour-blind women. It is worth noting, however, that very few biological sex-linked traits of this type appear to have any direct connection with psychological processes. This is partly because genetic psychological processes, if they exist at all, are likely to be *polygenic*. In human psychology it is the *phenotype* – the ever-developing outcome of the interaction between genetic and environmental influence – which is the focus of interest. See also *haemophilia, Y chromosome*.

Sex-role behaviour Behaviour which is influenced by the person's beliefs about what is appropriate for members of their own sex. The term can also be used to refer to behaviour which conforms to society's definition of appropriate gender behaviour.

Sex-role learning The processes by which a child or adolescent acquires an understanding of what is appropriate behaviour for their own sex, as opposed to appropriate behaviour for members of the other sex. Sex-role learning starts very early in life, and three-year olds have quite a clear idea of which gender related behaviours their parents think are appropriate.

Sex stereotypes Beliefs which are held in the *culture* about *sex differences* and appropriate *sex-role behaviour*. Like all *stereotypes* they make a useful starting point to know what to expect from a person, but easily become misleading if used in preference to observing what the person is actually like.

Sexism Discrimination against a person on the basis of their sex. It is often more subtle than *racism* because it is likely to be based on assumptions about *sex differences* which are widely held in society. As many of these assumptions have been developed to justify an unfair treatment of women (see *rationalization*), sexism is often taken to mean discrimination against women.

Sexual abuse A form of *child abuse* in which children are involved in inappropriate sexual activities, mostly with adults, and which is known to be psychologically damaging. It is now known that children of all ages, boys as well as girls, are sexually abused. Typical consequences involve distorting the child's ability to form appropriate relationships; limiting their ability to express affection in non-sexual ways; sometimes producing a high level of sensitivity to sexual cues, and a tendency to misinterpret ordinary interactions as sexual in content. Cases of child sexual abuse are sometimes detected by signs of unhappiness and an inability to concentrate at school. Since most cases of child sexual abuse involve incest the victims are often afraid of the consequences, and are therefore reluctant to disclose the abuse. Help for individuals who have been the victims of sexual abuse depends on the age at which it is identified, and may include *play therapy, family therapy,* individual *psychotherapy,* or self-help groups.

Sexual reproduction Forms of reproduction which depend on combining genetic material from a male and a female. The term is usually used in contrast to 'asexual reproduction' in which the offspring is produced entirely from genetic material provided by the single parent. Sexual reproduction has the major advantage of producing new combinations of genetic material and so increasing the diversity of the species. As the process requires cooperation between two members of the same species, it has resulted in the development of a great variety of interesting features such as courtship rituals, an ability to refrain from eating the sexual partner before their contribution to reproduction is complete, and biological motivations to ensure that the behaviour is undertaken however unlikely the required activities appear. Sexual reproduction is also widely regarded as being more fun than asexual reproduction.

Shadowing A task extensively used in studies of *selective attention.* Shadowing involves the audible repetition of a spoken message as it is received by the listener. In the classic experiments, subjects were presented with two messages simultaneously; one through each side of a pair of headphones (a *dichotic listening task*). They were asked to attend to only one of these messages, and in order to ensure that they were doing so, they would be required to 'shadow' the message. In this way, the effects of information input to both the non-attended and the attended ear could be assessed, as the spoken words of the subject would show what the subject was consciously noticing.

Sham rage An extreme form of uncontrolled rage, produced by direct action on the brain, usually electrical stimulation of the *limbic system*, and which ceases abruptly when the stimulation is switched off.

Shape constancy The perceptual adjustment which is made by the visual system when viewing objects from different angles, such that the *retinal image* varies. For instance, a cup seen from above casts a retinal image very different from that of a cup seen from the side; yet is perceived as having a constant shape. See also *size constancy, colour constancy.* [f.]

Shape constancy

Shaping See *behaviour shaping*.

Shock therapy See *electroconvulsive therapy*.

Short-term memory (STM) Memory which lasts for only a few seconds, like the kind of memory used when retaining a telephone number while dialling. The concept of short-term memory was first introduced by William James in 1890, and has been used extensively in psychological theories of memory ever since. One of its notable characteristics is its vulnerability either through a rapid decay of the memory trace, or displacement by new material. This means that in order to retain material for any length of time, it is necessary to rehearse it continuously. Another characteristic is limited capacity, with old information being displaced to make room for new. This limited memory was identified by Miller as consisting of 7 items ± 2; but the amount of information contained in those 7 items could be extending by chunking information into meaningful larger units. Some theorists, notably Atkinson and Shiffrin, see short-term memory as an initial stage for material entering *long-term memory*; although they also see it as a completely different type of memory store. In recent years, the *levels of processing* approach to memory has implied that the existence of two separate memory stores is an unnecessary refinement, and that the characteristics of STM can be seen simply as the effects of the very superficial processing which information receives when first perceived.

Sibling A word used to refer to a brother or sister, which has the advantage of not denoting the gender of the person being referred to.

Sibling rivalry The commonly observed jealousy between siblings, which may start from a competition for attention and affection from the parents but then generalizes to many or all aspects of their lives.

SIDS See *sudden infant death syndrome*.

Signal-detectability theory A theory about how weak signals are detected despite the presence of background *noise*. By making simplifying assumptions, (in particular that only the level of noise and the level of signal are to be considered, and that when both are present the levels simply add to the total sensation, rather than interacting or cancelling each other out) it has been possible to produce a mathematical analysis of the process of detecting signals. This approach has been effective in certain restricted cases, and much of the theory is incorporated in the *receiver-operating characteristic curve*.

Signal-detection task A task used to investigate how long a subject can perform effectively, when asked to identify one particular type of signal appearing at random intervals amid other distracting stimuli. The task might be auditory: a tone lasting a bit longer than other tones which are sounded at intervals, for instance; or it might be visual, such as the detection of one special shape appearing among other shapes. Some signal detection tasks are replications of the displays which a radar operator would scan, allowing researchers to identify potential sources of error, and to investigate possible alleviating measures. See also *sustained attention*.

Simple cell A type of neurone found in the *lateral geniculate nuclei* of the *thalamus* and also in the *visual cortex*, which will fire only when a very specific stimulus occurs within the *visual field*. First identified by Hubel and Wiesel, simple cells will respond either to a particular dot or line in a specific part of the visual field, or to a line at a particular orientation in any part of the visual field. There is some evidence also that something like 90% of these cells can adapt their functioning, if visual experience is limited. After a *critical period*, their functioning becomes relatively fixed. It is thought possible that disorders of the arrangements of simple or *complex cells* may produce astigmatism. See also *hypercomplex cells*.

Simulation Any process of modelling or imitating an actual, real-life event. The term is used in psychology to refer to: apparatus which mimics a real situation in which training can be more safely carried out, as in aeroplane cockpit simulators; people who act as if they have psychological or physical conditions, as in faking epileptic seizures; and in *computer simulation*.

Simultaneous conditioning A variant of *classical conditioning* in which the *unconditioned stimulus* is presented at exactly the same time as the *conditioned stimulus*. See also *trace conditioning, delayed conditioning*.

Single-blind control An experimental control in which the subjects in a study are unaware of the hypothesis which is being investigated, but the researcher does know it. See also *double-blind control*.

SIT See *social identification theory*.

Situational attribution In *attribution theory*, this refers to explaining a person's behaviour or experiences as arising from the situation that they are in rather than from the personality or other internal characteristic of that person (which would be a *dispositional attribution*). See also *attributional error*.

Size constancy The perceptual process by which objects are judged to be consistent in size, regardless of the actual dimensions of the image which they cast on the *retina* of the eye. An object viewed from a distance will produce a *retinal image* which is very different in size from the same object seen at close quarters, but the perceptual system adjusts its recognition of the object, such that in both cases the size is seen as being the same. In extreme conditions size constancy may break down, such as when cars or people are viewed from the top of a skyscraper. See also *shape constancy, colour constancy*.

Skewed distribution curve A version of the *normal distribution curve* which is not symmetrical, in that one side is extended further than another. For example: a curve plotted from measurements of simple reaction times will be skewed, because while there is a physiological limit to how quickly someone can react to the stimulus, there is no limit to how long they can take. So a curve drawn from such measures will tend to 'lean' towards the left, but have a 'tail' which stretches out to the right. This is known as a positive skew; a curve which 'leans' in the other direction is referred to as negatively skewed. See also *measures of central tendency*.

Skinner box A device developed by B.F. Skinner for investigating operant conditioning. A typical Skinner Box will contain little other than a lever, a food-delivery chute, and a signal light. When a hungry small animal such as a laboratory rat is placed in the box, its exploratory behaviour eventually results in its pressing the lever; at which point a food pellet is delivered. This *reinforces* the lever-pressing action, rendering the animal more likely to repeat it. This process results in the learning of lever-pressing as a means of obtaining food, although the experience of one of the authors suggests that this only happens if the animal feels inclined to cooperate, and is not inevitable. The preliminary phase of getting the animal to push the lever for the first time will be quicker if a *behaviour shaping* procedure is employed. The signal light can be used as a *discriminatory stimulus*; and the Skinner Box may be set to deliver *partial reinforcement* according to a *reinforcement schedule*. [f.]

Sleep cycles Patterns of sleeping which involve changes in *EEG recordings* produced by a sleeper, and corresponding differences in how easy the person finds it to wake up. During a typical night, sleepers pass through the different levels of sleep in a cyclic fashion between five and seven times. Levels 1 and 2 are light

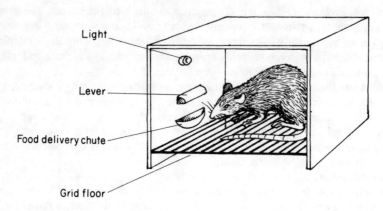

A skinner box

sleep characterized by irregular EEG patterns; the deeper levels 3 and 4 show regular wave patterns in EEG recordings. Typically, the sleeper will cycle through the levels every 40 to 80 minutes, and then enter REM sleep for a period before starting a new cycle. During a period of normal sleep, deeper stages become shorter and then cease completely, while the REM stage becomes longer. See also *rapid eye movement sleep; orthodox sleep.* [f.]

Sleeper effect An experimental effect which is not apparent immediately but which may appear later. For example an item might be stored in memory but not be accessible on testing soon after the acquisition. However it may be recalled the next day.

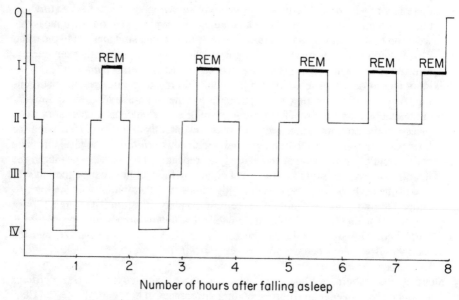

Cycles of sleep levels during the course of a typical night

Sociability The ability to engage in an appropriate range of social relationships. The different forms that sociability takes at different ages and the means by which it develops is one of the major topics of *developmental psychology*.

Social attribution A branch of *attribution* theory which attempts to integrate the social orientations of *European social psychology* with analysis of the nature of individual everyday explanation. Where more traditional versions of attribution theory, such as the *covariance* approach, have treated attribution as the product of individual cognitive processes, social attribution emphasizes the social nature of many of the explanations adopted by people, and tends to focus more on intergroup similarities and differences than on individual problem-solving. See also *social representations, social identification theory, lay epistemology*.

Social class The classification of people according to their occupations and economic circumstances. Naturally, such a classification generates all kinds of problems but the findings of widespread differences between different classes is consistent enough to motivate researchers to continue to divide people in this way. It is important to recognize that social class, in itself, cannot be an explanation of anything, though it is often used in this way. A problem in interpreting social class differences is that since the different classes differ in almost every way possible (education, income, health, smoking, religious attitudes, etc.) it is not usually possible to say what any particular difference is caused by. The most commonly used criterion for allocating social class in Britain is called The Registrar General's Classification and consists of a list of occupations allocated into classes from 1 to 5. The term socio-economic status is sometimes used in an attempt to avoid the undesirable implications of *class*.

Social cognition (i) The branch of social psychology concerned with people's understanding of what is going on in social interaction. This tends to be mainly concerned with identifying the different forms of social assumptions and social explanation. Social cognition therefore includes the study of *social schemas* and *scripts*, as well as *social representations* and *social attribution*.

(ii) In *developmental psychology*, it is an approach to cognitive development which states that social interaction is the most important factor in the development of the young child's cognition. Work in this field has produced some re-evaluation of the classic Piagetian findings concerning *conservation* and *egocentricity*, as it appears that the traditional responses obtained from children were more a product of the child's interpretation of the social demands of the experimental situation (saying what the experimenter wanted to hear, etc.) than with any inability to conserve or decentre on the part of the child. Work by Judy Dunn and others indicates that children show cognitive abilities within social interactions at much earlier ages than they can show them in the context of physical science, which was the basis of Piaget's investigations. Other studies have examined the influence of *social expectation* and *modelling* in cognitive development.

Social comparison Social comparison is concerned with the way in which we automatically draw comparisons between different groups and individuals. Social comparison leads to a number of outcomes, including *social identification*, as people assess the relative status, power etc of their own group relative to others. Festinger also proposed that social comparison leads to a tendency towards shared beliefs, particularly with respect to social judgements. In the case of beliefs about the physical world, beliefs can be directly tested: we can observe directly that glass is fragile by breaking some. But in the case of social beliefs, such as whether, say, a socialist form of government leads to greater prosperity, we have no such access to direct factual information, and consequently will come to depend more on the views of others. Here social comparison comes into play, as we will be more likely to accept the views of those we consider to be similar to ourselves than of those we see as different.

Social constructionism The position taken by some social psychologists that social reality is constructed between people, rather than being an objective phenomenon of which there can only be one true description.

Social exchange theory The idea that social functioning operates according to a basic rule that people should benefit from a social exchange to about the same extent as they have contributed to it. See *equity theory*.

Social facilitation The finding that performance is usually improved by the presence of others. Simple and well rehearsed tasks are most likely to be facilitated, so if the presence of others is a source of arousal, the phenomenon follows the *Yerkes–Dodson law*.

Social identity theory (SIT) The theory developed particularly by Tajfel (e.g. Tajfel, 1982), whic proposes that the membership of social groups, rather than being some kind of external act, or role, actually forms a highly significant part of the *self-concept*. Social identification theory draws on two fundamental psychological mechanisms. The first of these is the *cognitive* mechanism of classification, whereby objects, events and people are classified into categories, and compared with one another. The second is the tendency for people to seek sources of positive *self-esteem*. The outcome of these two processes is social identification, as the tendency to classify also leads people to classify their own groups, and to compare the social groups to which they belong with others. If their group membership provides a source of positive self-esteem, the individual will come to identify strongly with the group, and to incorporate group membership as part of their self-image. If such comparisons do not reflect positively on the self-concept, the individual will seek to leave the group (social mobility); to distance themselves from it; or to alter the perceived status of the group to which they belong (social change). Social identification may also lead to the emergence of shared beliefs, or *social representations*, within a given group.

Social identity theory forms a core theory in the school of thought known as *European Social Psychology*. This school is particularly distinguished from the majority of social psychological theories in its emphasis on the realities of social life in terms of differences in social status, relative power and access to economic resources. Other theories of this school include social representation theory and some versions of *attribution* theory. See also *social comparison* and *minimal group paradigm*.

Social impact theory An American social psychological theory proposed by Latané (1981), in which the strength of social impact in phenomena such as conformity is perceived as increasing with the number, immediacy and strength of the sources. In other words, social impact, or social pressure, is higher if there are more people exerting it; if those people are closer to the individual rather than distant; and if they are important people rather than simply random strangers. The second aspect of the theory concerns diffusion of impact, proposing that the strength, or influence, of a source decreases as the number, immediacy and importance of the targets towards which it is directed.

Social impact theory has been hailed by some social psychologists as providing a higher order model which can account for a number of diverse findings in social psychology. But it has also been sharply criticised for its *reductionist* approach, in that it sees social influences simply as the product of the actions of individuals, and fails to take account either of *emergent properties* of social groups, or the importance of social contexts. It therefore represents a direct contrast to the school of thought in social psychology known as *European social psychology*.

Social interaction A process in which two people or animals directly influence each others' behaviour. Social interaction is the core phenomenon of *social psychology*, and the complex regulation of forms of social interaction is an important part of the young child's *socialization*.

Social learning theory An approach to child development which states that children develop through learning from the other people around them. Social learning theory emphasizes the processes by which children come to adopt the rules, norms and assumptions of their society, e.g. *operant conditioning, imitation*, and *identification*. In general, social development is seen as a continuous learning process, rather than as happening in stages and many theorists consider it to continue throughout adult life.

Social needs The third level in Maslow's *hierarchy of needs* is concerned with group identity and membership, love, and positive interaction with others. According to Maslow, social needs become important once the basic *physiological needs* and the *safety needs* have been satisfied. Once the social needs have been adequately met, *aesthetic* needs become important. At the top of the hierarchy is *self-actualization*, which Maslow considers to be possible only once all other levels of need are satisfied. Many psychologists criticize this model of human needs on the grounds that it does not account for many instances of human behaviour in which 'higher' needs are apparently put before basic ones, the classic example being the case of the 'starving poet'; there are also many examples of *prosocial behaviour* in the face of physical deprivation.

Social norms Forms of behaviour which are widespread within a society and/or are widely accepted as appropriate. Often, it is the second condition which is more important: for example, there are probably far more people in our society who abuse children than the number who work professionally for their welfare and protection. Yet concern for children, rather than abuse of them, is still accepted as the norm. Acceptance of a person in a society is usually based on the extent to which the person follows, or at least expresses agreement with, the norms.

Social psychology The branch of psychology which is particularly concerned with the nature and form of social interaction and how people come to influence each others' behaviour. As such, it includes the study of social phenomena, such as *conformity, obedience* and *non-verbal communication*, as well as aspects of social cognition such as *social perception, attitudes* and *attribution*. In recent years, a distinction has been developing between the problem-centred and individualistic (some say *reductionist*) approaches to the understanding of social phenomena seen as particularly typical of American social psychology, and the group-based, highly contextual form of social psychology which has become known as *European social psychology*, and which includes the study of *social identification, social representations*, and *social attribution*.

Social representations A concept developed and articulated by Moscovici (e.g. Moscovici, 1984), social representations are the shared beliefs adopted by groups of people, and used to explain social experience. Social representations vary in scope from the large-scale ideological beliefs shared by a society in general, to smaller scale beliefs adopted by members of a specific social group or subculture. Despite their shared nature, however, social representations are dynamic, negotiated through social interaction and conversation, and modified or adapted as they become incorporated into the world-knowledge of the individual. One of the major contributions to the group of theories known as *European social psychology*, social representations act as the cognitive interface between individual action and social ideology, and have been studied in terms of several social movements, including changes in health and dietary beliefs over time. See also *lay epistemology*, and *social attribution*

Social role See *role*.

Social schema A form of schema which is particularly concerned with social cognition and social interaction. As with other forms of *schema*, the social schema serves not just to assimilate and interpret experience, but also to direct action. A number of different types of social schema have been identified, among them *scripts*, the role-schema, which is particularly concerned with the social roles to be placed in society, and the person-schema, which is concerned with structuring and applying knowledge about people. In view of the overwhelming evidence as to the importance of social factors in the development of the self-concept, the self-schema has also been identified as a type of social schema.

Social sciences A collective term for those academic disciplines which involve the study of human beings interacting with one another. As such, it includes *psychology, sociology, anthropology, linguistics,* economics, geography, etc.

Socialization The process by which a child becomes integrated into society by adopting its *norms* and values, acquiring the necessary skills of *social interaction*, and learning to adopt an acceptable *role*.

Sociobiology A *reductionist* approach to the study of social behaviour, in which the identification of a 'unit of natural selection' which could possibly form the basis for a social phenomenon is taken as an 'explanation' for the phenomenon. The 'unit of natural selection' is referred to as a 'gene', although it is not biologically equivalent to the *gene* as studied by geneticists. All behaviour is seen as being directed towards the perpetuation and replication of genes. Even *altruistic behaviour* is interpreted in terms of the perpetuation of the 'selfish gene', through the mechanism known as *kin selection*.

There are many weaknesses in the sociobiological approach, one of which is its retrospective approach to methodology, in which explanation takes three stages: (1) the identification of some kind of 'universal' in behaviour; (2) the identification of a possible 'unit of natural selection' which could produce such behaviour; and (3) the development of a plausible account of how that behaviour could (or, more often, 'must') have evolved. Other objections stem from the highly selective approach both to the 'universals' of behaviour – which usually emphasize only the more negative human traits – and those examples of animal behaviour taken as evidence, in which behavioural variations are largely ignored, and only those cases which support the argument are acknowledged.

Socioeconomic status or background An elaborate way of referring to *social class* while attempting to avoid the unwanted implications and problems of definition and distinctions involved in the concept of class.

Sociolinguistics The study of social forms of language, and the ways in which language is used in society. Sociolinguistics inevitably shows considerable overlap with, and can make contributions to, *social psychology*, for instance in the study of the social influence of accents and dialect and in the study of *elaborated* and *restricted codes* of language.

Sociology The systematic study of societies and other social institutions, their effects on people and how people operate within them. There is much overlap between sociology and *social psychology*.

Somatic nervous system The network of nerve fibres which carries messages from around the body to and from the *central nervous system*. 'Somatic' means 'of the body', and this nervous system consists mainly of *sensory* and *motor neurones* throughout the body, linked by the *spinal cord* and the *brain*. This allows for bodily sensations, movement and experience to be recognized in the central processing areas of the nervous system. See also *autonomic nervous system*.

Somatosensory area A strip running alongside the *central fissure*, in the *parietal lobe* of the *cerebral cortex*. This area is the *sensory projection area* which is particularly concerned with the sensation of touch. Different parts of the somatosensory area correspond to different areas of the body; those parts of the body which are more sensitive have a correspondingly greater amount of surface area on this strip.

Somatotype An overall bodily shape, which has been thought by some researchers to provide an indication of personality. One of the most famous researchers in this area was Sheldon, who classified human bodies into three main groups: ectomorphs, with a tall, slender physique, endomorphs, who were plump and rounded in shape and mesomorphs, who were sturdy and muscular. Sheldon saw this as indicative of *personality*, considering that ectomorphs tended to be introverted and were often nervous and intellectual types, endomorphs tended to be friendly and relaxed people, while mesomorphs tended to be noisy, hearty, and

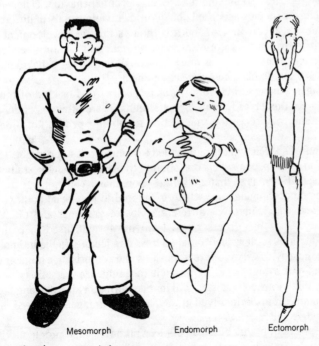

Mesomorph   Endomorph   Ectomorph

The three somatotypes

often callous in their interpersonal interactions. Although Sheldon's studies involved an impressive sample size, they were methodologically flawed, and took little account of *experimenter bias* or of *self-fulfilling prophecies*. [f.]

Spaced practice See *distributed practice*.

Spastic Affected by muscular spasms. The term has been used to refer to people suffering from *cerebral palsy* but as it has entered the common language as a derogatory label for clumsiness it is rarely used within psychology.

Spearman's rank–order correlation coefficient A measure of *correlation* which can be applied to *ordinal data* and which is usually used for small *samples*. In the event of it being used for a large sample (e.g. over 60 pairs of scores), the final coefficient obtained from the test is considered to be equivalent to a *Pearson's product–moment correlation coefficient*. See also *scattergram*.

Special child A term adopted to refer to all children whose qualities or abilities are well outside the normal range. It represents an attempt to avoid the automatically negative implications of terms like *mentally handicapped* and retarded, and to make an association between children who need special resources because of some disadvantage and those who need special attention because they are exceptionally gifted in some way. More recently, the expression 'children with special needs' has been adopted to reduce the possibility of *labelling*.

Special learning difficulties See *mental handicap*.

Species-specific behaviour Behaviour which occurs in all members of a given species and which does not appear to take place in animals of other, even closely related, species. One obvious example is language in humans. Whether or not one believes that other primates can be taught a language, the fact remains that only humans develop this complex means of communication spontaneously, and it occurs in all human societies, so making it a species-specific behaviour. There are many other examples of species-specific behaviour. Courtship rituals in different species have been extensively studied, and it is thought that the development of elaborate mating patterns serves to prevent inappropriate cross-mating between members of similar species. It is usually assumed that if a behaviour is species-specific it is likely to have an *innate* component.

Specific hunger Hunger which is directed towards a specific food or kind of food, e.g. a hunger for sweet things, or for salt. Specific hungers are often experienced during pregnancy, and may serve the function of supplying specific nutritional needs.

Speech acts Segments of speech which are intended to bring about some effect. The focusing of attention onto speech acts is one attempt to narrow down the study of *language* to more specific areas so that it becomes more manageable.

Speech register A mode of language use which is tailored to the social context in which it is used, and which involves different styles of grammar and often a different vocabulary. Speech registers range from the formal, used in highly structured social situations such as an official address or a lecture, to the intimate, used only between those with very close relationships and comprising a number of shared assumptions and a high level of implicit meaning. Conversations with friends, using an affiliative speech register, will involve different kinds of language use from the consultative register involved in, say, asking a stranger for directions. See also *accent, dialect* and *psycholinguistics*.

Speech therapy The profession which helps people who have some problem with verbal communication. Speech therapists use many techniques from psychology,

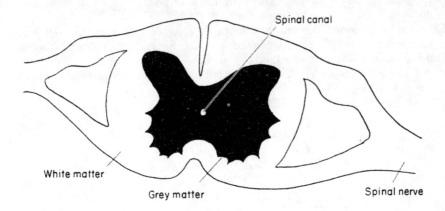

A cross-section through the spinal cord

particularly behavioural methods, and are increasingly paying attention to social factors in the disruption of communication.

Spinal cord The bundle of nerve fibres which runs up the channel within the spinal cord. The spinal cord forms a pathway between the *somatic nervous system* and the brain, mediates some basic functions such as pain reflexes and, in the higher segments, some of the functioning of the *autonomic nervous system*. In cross-section, the spinal cord can be shown to consist of an outer layer of *white matter*, with an inner part of *grey matter*, and a small central canal at the core, which contains cerebro-spinal fluid. As the spinal cord is the medium by which the brain transmits information to the body, lesions of the spinal cord can result in paralysis; the extent of the paralysis depending on how far up the spinal cord the lesion occurs, lesions closer to the brain producing a more total paralysis. [f.]

Split-brain studies Studies of subjects in whom the corpus callosum and the optic chiasma are severed. Originally resulting from an operation on humans as an attempt to control severe epilepsy, the condition was found to permit the study of the independent functioning of the two *cerebral hemispheres*. This work extended knowledge of localization of function in the brain, for example that logical/mathematical functioning is localized in the left hemisphere, while artistic abilities and spatial awareness were more highly developed in the right hemisphere. It also led to the discovery that the two halves of the brain could operate independently as decision-making and intelligent structures.

Split-half reliability A technique for assessing the *reliability* of a test by calculating a score from first one half of the items, then from the other half to see whether the two scores agree.

Split personality See *multiple personality*.

Split-span tests Tests first developed by Broadbent to study *selective attention* in which a succession of digits is presented to a subject through headphones, in such a way that two different digits are presented simultaneously, one to each ear. Broadbent's observation was that, when asked to repeat the digits they had heard, subjects did not mix digits from different ears, but instead reported a succession from one ear only or from each ear in turn, thus implying a 'filtering' approach to attention.

Spontaneous recovery The sudden reappearance of a learned response after it has been *extinguished* due to lack of reinforcement. Spontaneous recovery has been demonstrated in both *operant* and *classical conditioning*; and, if the spontaneous response is reinforced, can lead to the reappearance of the learned behaviour at full strength, very quickly.

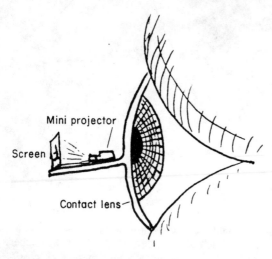

Creating a stabilized retinal image

Stabilized retinal image The finding that *rod* and *cone cells* in the *retina habituate* quickly if they are exposed to a stable image. This does not normally occur because of *saccades*, but has been achieved experimentally by the use of small projectors and screens attached to contact lenses. The effect of maintaining the image of an object at a fixed position on the retina is that the subject ceases to be able to see it. [f.]

Stable attributions *Attributions* in which the cause which has been identified is of a kind that will apply again in similar situations in the future. For example, believing that you have failed a summer exam because of your hayfever is a stable attribution, in that you are likely also to have hayfever for future summer exams.

Stage Many theories in developmental psychology are based on the concept of development from stage to stage. Major examples are Freud's *psychosexual stages*, Gesell's *maturational* stages, Piaget's cognitive stages, Erikson's stages through the life span and Kohlberg's stages of *moral development*. In all cases it is assumed that each stage must be completed more or less successfully before the next stage can be adequately tackled. This entails that stages will occur in a fixed order since later stages depend on earlier ones. The theories differ in whether they see the transition from one stage to the next as gradual or abrupt, and in what happens to the earlier stages. Some, like Kohlberg's, assume the earlier stage becomes irrelevant and is abandoned once a new stage is reached. Freud sees the earlier stage as something to be relinquished if possible but likely to continue to exert an influence. Piaget and Gesell see earlier stages as built on and incorporated into later functioning, but no longer used in their original form. Another developmental stage theorist, Heinz Werner, sees earlier stages as more primitive modes of functioning which may still have their uses in certain circumstances and which can still be used when the occasion arises, a rather more positive view of the process that Freud identified as regression. Broadly, stage theories imply qualitative differences in functioning at different ages and can be contrasted with behavioural approaches, such as *social learning theory*, which assume that the same processes apply throughout the life span.

Standard deviation (SD) A statistical *measure of dispersion* of a set of scores. Calculation of the standard deviation is a basic step in *parametric tests*. Simply knowing that two scores are 5 points apart tells you nothing unless you know how widely dispersed the scores in the population are. If the SD is 1, then our difference of 5 points indicates a wide divergence on what is being measured. If the SD were 100, 5 points is no real difference at all. The SD is the square root of the *variance*. [f.]

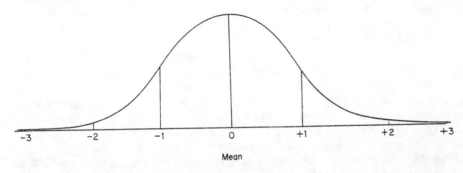

Standard deviations of the normal distribution curve

Standardization Establishing a set of standardized *procedures* for a test which ensures that results are comparable when obtained in different settings. The term can refer either to procedures for administering the test, or to data which indicate the expected range of scores in specified populations (*norms*).

Standardized instructions A pre-determined set of instructions which is given in the same manner and using the same words to each subject taking part in a given experimental procedure. The use of standardized instructions is intended to provide a control against *experimenter effects* in research or testing.

Standardized procedures A set of experimental procedures, or sequence of events, which has been established in advance, such that it will be carried out in the same way for each subject. This is one way of controlling unconscious *experimenter effects* from influencing the results of a study.

Stanford–Binet test The most extensively used *intelligence test*, developed at Stanford University in 1916, using Binet's test as a starting point. The test has been revised several times but retains a major limitation in only giving a single *intelligence quotient* score. Later tests such as the *Wechsler Adult Intelligence Scale* and the *British Abilities Scale* provide independent measures of various aspects of intelligence.

State-dependent learning Learning which is demonstrated most effectively when the individual is in the same physiological state as when it was originally learned. For instance, information learned when a subject is under the influence of alcohol is often most readily recalled at times when the subject is again under the influence. State-dependent learning may similarly be demonstrated with a range of drugs, including *amphetamines* and *tranquillizers*.

Statistic A measure of some aspect of a sample. Measures of a population are called *parameters*.

Statistics The set of mathematical techniques designed to summarize raw data and indicate the conclusions that can be drawn. Statistical techniques are largely concerned with determining whether a given result could easily have been obtained by chance or whether it would be better to assume that some consistent effect was operating. See *hypothesis testing; parametric statistics; non-parametric statistics*.

Statistical significance A statement of how likely it is that the outcome of a study or comparison has simply occurred through chance factors. Statistical significance is usually expressed in terms of a ratio of 'p'; e.g. 'p is less than (or equal to) 0.05'; or 'p is less than (or equal to) 0.01'. In such expressions, 'p' stands for the probability that the *null hypothesis* is correct, in other words, the probability that the results have simply occurred by chance. Before a study is conducted, the level of significance considered to be acceptable to the researcher will have been decided: $p \leq 0.05$ means that there is only a 5% chance (or less) of the null hypothesis being correct; and this may be acceptable to the researcher. Alternatively, in a study with important social or ethical implications, such as the testing out of a new drug, a far more stringent level of significance might be required, such as $p \leq 0.0001$, and this too will have been decided in advance. Given the highly variable nature of the subject matter in psychology, the concept of statistical significance is at the core of most psychological research. See also *Type 1 error*.

Stereoscope A device much used by early investigators of perception, which allows a researcher to present two different pictures to a subject simultaneously: one to each eye.

Stereoscopic vision Vision which provides a direct perception of depth or of a 3-dimensional image. It is achieved by integrating information concerning the same visual field, received through two eyes simultaneously. The cortex relates the information from equivalent parts of the retina, which will be receiving slightly different patterns from the same source because of the distance between the eyes, and uses the differences to construct stereoscopic vision. This can only occur in animals with frontally-mounted eyes, such as humans and other primates, cats, owls, etc., and cannot take place in animals like rabbits or blackbirds, with eyes at the side of the head. Stereoscopic vision is particularly useful for the accurate judging of distance, through the process of *binocular disparity*; and it is thought that this may provide an evolutionary explanation for its development in the largely arboreal (living in trees) primate group.

Stereotype A belief about a class of people which is then applied to individual members of the class to provide expectations about the person in the absence of specific knowledge. Stereotypes enable us to begin interaction with strangers with an expectation of better than chance success in choosing an appropriate style and topic of conversation. Stereotypes can therefore be seen as highly functional in a setting which involves frequent interactions with people of whom one has limited knowledge. The view of stereotypes as undesirable arises from assuming either that they will be inaccurate or that they will persist despite contrary information.

Neither assumption is necessarily true. If a stereotype is inaccurate, negative, and held to despite contrary information, it qualifies to be called a *prejudice*.

Stimulants Drugs which produce heightened activity of the *central nervous system*, often used to combat fatigue or tedium. The most commonly used stimulant is probably *caffeine*, which is consumed daily in the form of tea, coffee, or cola by many people worldwide. In medical use, *amphetamines* are one of the most common groups of stimulants, and are also used as *recreational drugs* for the same purpose, as is *cocaine*. One of the more common uses of amphetamines is as appetite suppressors; and many other stimulants appear to have similar properties, although to a lesser degree.

Stimulus Any event to which an organism, human, animal or plant, responds. Stimulus is a general term which avoids specifying the form in which stimulation is presented; it essentially refers to anything which is detected by the sensory equipment possessed by the organism.

Stimulus discrimination The form of discrimination shown in stimulus–response learning, in which a response will occur to one specific stimulus, but will not occur in the presence of a similar stimulus. Unlike *stimulus generalization*, which occurs without prior training, stimulus discrimination is learned by the organism through *reinforcement*. Responses made in the presence of one stimulus are reinforced, those made to the other are not. In this way, the organism comes to discriminate between the two.

Stimulus generalization When a learned response is produced to a stimulus different from the one to which it was originally learned. Stimulus generalization often shows the *generalization gradient* that the response is strongest to those stimuli which are most similar to the original one.

Stimulus–response learning Learning which occurs as a result of the association between a *stimulus* and some kind of behavioural response. In general, there are thought to be two basic forms of stimulus–response learning: *classical conditioning* and *operant conditioning*. Some psychologists classify *one-trial learning*, in which such an association is formed as a result of only one learning trial or experience, as a third form; others regard it as a special form of classical conditioning.

STM See *short-term memory*.

Strange situation technique A standardized method developed by Mary Ainsworth (e.g. Ainsworth, 1979) to study attachments in one-year-old children. The child is brought into an unfamiliar environment by its mother, then a stranger enters and mother leaves. Finally, mother returns. The reactions of the child are recorded in a standard way, and the quality of the attachment is judged. Ainsworth classified attachments as either secure (Type B), anxious (Type A) or ambivalent (Type C). The technique has made it possible to study the consequences of these different forms of *attachment*.

Stratified sampling A technique of collecting a *sample* which is designed to make the sample represent as accurately as possible the *population* from which it was recruited. The major groupings, e.g. social class, in the population are identified, and the sample is recruited from each of these groupings, so that each can be analysed separately if need be. See also *quota sampling, random sampling, opportunity sampling*.

Stress Usually, the effect on a person of being subjected to noxious stimulation, or the threat of such stimulation, particularly when they are unable to avoid or terminate the condition. Major changes in one's life (*life events*) have been found to be a common source of stress which leave people vulnerable to *depression*. Hans Selye has found similar physiological and psychological reactions to prolonged stress regardless of the nature of the source (see *general adaptation syndrome*). While stress is unpleasant and often damaging it is also recognized that it may be actively sought (as when apparently sane people jump out of airplanes for fun), and is an important source of motivation. The term is also sometimes used for the source of the stress (noise, poor housing, etc.) but it would be better if such conditions were always called 'stressors'.

Striate cortex See *visual cortex*.

Stroboscopic motion The phenomenon, which forms the basis of film projection, that a series of separate pictures, shown in rapid succession, will seem to produce a continuous movement. Stroboscopic motion can also be demonstrated using lights which flicker on and off, as in the *phi phenomenon*, and takes its name from the brief appearance of each image, in the same way that a stroboscope (a light which flashes rapidly on and off) produces a succession of 'flash pictures'. See also *movement illusions*.

Stroop effect A reliable experimental effect which demonstrates how powerful routine cognitive processing can be. The Stroop effect is normally demonstrated using colour names: two sets of different colour names (orange, red, blue, etc.) are written on cards: one set is written in the appropriate colour for the word, while the other set is written in a different colour (e.g. 'orange' written in green ink). On being asked to identify the colours in each list, subjects take longer to process the information on the cards which contain the discrepant information. Reading the colour name occurs as an automatic cognitive sub-routine, which interferes with the recognition of the colour itself.

Structuralism An approach to theory in which psychological phenomena are explained as the natural outcome of the way the organism is structured. The proposed structures may be physical and open to direct examination (e.g. accounts of aggression based on interpreting brain structure) or hypothetical. Examples of the latter are Freud's personality structure and Piaget's cognitive structures. Structuralist approaches in anthropology and sociology are concerned with the social structures within which people function, though these are often taken to be outward manifestations of mental structures. The term is also applied to attempts to understand how language works by examining its structure. Structural theories are contrasted with *functional* approaches.

Study skills The set of techniques, strategies and behaviour patterns which form a structured approach to learning; often based on psychological theory, but also on experiences acquired and transmitted less formally. Study skills can be related to the theoretical area of *metacognition*, but is usually treated as a separate topic in its own right. It includes such features of effective study as reading skills, effective revision techniques, organizing study time, and examination strategies.

Subconscious Material of which the person is not consciously aware but which could be made conscious if required. The term has the same meaning as *preconscious* but is not so strictly tied to Freudian theory.

Sub-cortical structures Those parts of the brain which are found below the *cerebrum* – in other words, all the parts of the brain except for the *cerebral hemispheres*. [f.]

Sublimation In Freudian theory, the redirection of instinctual energies towards more socially acceptable goals. During development, direct expression of psychosexual

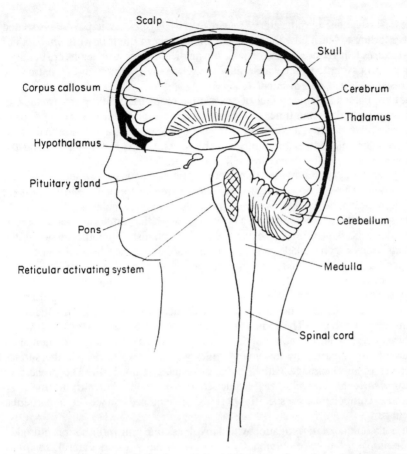

Scalp

Skull

Corpus callosum

Cerebrum

Thalamus

Hypothalamus

Pituitary gland

Pons

Cerebellum

Reticular activating system

Medulla

Spinal cord

A cross-section through the human brain showing the sub-cortical structures

drives is prohibited and the energies are diverted into substitute activities which are more acceptable. In this way society's restrictions on the direct gratification of instinctual needs results in energy being made available for purposes that are valued by society. A more general term is *displacement*.

Subliminal perception Perception which occurs in such a way that the person is unconscious of it. Several studies have demonstrated that information may be absorbed by the perceptual system extremely rapidly, and in such a way that it does not penetrate to conscious awareness, but may nonetheless influence people at an unconscious level. Studies which involved presenting threatening or offensive stimuli subliminally have demonstrated marked alteration in the subject's *arousal* level as a consequence. Subliminal advertising is prohibited in the U.K. by the Broadcasting Acts; but is permitted in private locations, such as supermarkets, providing a notice is displayed informing the public that this is occurring. In such cases, it normally takes the form of faint auditory messages embedded in music.

Successive approximation See *behaviour shaping*.

Sudden infant death syndrome (SIDS) Also called cot death or crib death. Babies appear to go through a vulnerable period around 2 to 4 months and during this time a significant number are found dead in their cots, having shown little or no sign of illness or any other warning sign. Some research suggests that it may be associated with a failure to learn how to restart breathing early in life following *apnoea*, but most research has concentrated on possible medical causes. Cot deaths are of major concern to psychologists because they are relatively common and an extremely distressing form of *bereavement*.

Summation The cumulative effect of several neurones transmitting information to one neurone at the same time. If a single *synaptic transmission* is received, from one other neurone only, it is unlikely to be enough to produce a response in the next cell. But the total effect brought about by several *receptor sites* receiving the *neurotransmitter* at the same time will produce the effect.

Superego In Freudian theory, the third component of personality which forms after the *id* and *ego* have become established. The superego is formed in early childhood by internalizing the parents' system of rewards and punishments, so that the child comes to operate according to these rules even when the parents are not present. It is not quite the same thing as the conscience as it retains an infantile version of the parents' rules which is likely to be severe and intolerant. The adult conscience may be much more realistic and sophisticated and so may come into conflict with the superego.

Surface structure The term coined by Chomsky to refer to the pattern of grammar and sentence structure which are found in a particular language, and distinguish it from other languages. The term is used in contrast with *deep structure*, which, Chomsky argues, is common to all languages and which forms the fundamental set of principles, inherited by the young child, which it uses to decode the surface structure of the language which it hears around it from birth. The process of *transformational grammar* was developed as a method of identifying the deep structure components of specific phrases or sentences made in a particular language.

Survey A technique of investigation which involves collecting information, attitudes, or opinions from large numbers of people, usually using careful *sampling* procedures. Although a survey rarely allows for in-depth investigation of a topic, it can be extremely valuable for investigating general patterns of human behaviour, such as surveys of sleeping habits or attitudes.

Sustained attention Also referred to as vigilance in many accounts, this refers to an extended period of concentration on a relatively simple task. Studies of sustained attention became important during the Second World War, with the development of a defence technology, since errors brought on by fatigue or distraction could have momentous effects, especially in the case of radar surveillance. Overall, studies of sustained attention have tended to take the form of *signal-detection tasks*. Performance on these has been shown to be positively affected by such variables as the presence of others, a limited amount of extraneous noise, a high degree of *introversion* in the individual concerned, etc. One theoretical explanation which has been suggested is that all of these relate to the degree of *arousal* experienced by the subject as they are carrying out the task.

Symbolic representation The third of the *modes of representation* described by Bruner, in which information is stored as symbols, such as numbers, words, or signs. Bruner argued that this mode of representation enables the child to organize and categorize information, and to perceive relationships which might not otherwise have been readily identifiable. As such, he regards the development of symbolic representation, especially through the use of language, as being of paramount importance in cognitive development. See also *enactive representation, iconic representation.*

Sympathetic division One of the two divisions of the *autonomic nervous system*, the sympathetic division is the set of nerve fibres which, when stimulated, increase *arousal* and may trigger off the *fight or flight response*, producing a rapid burst of energy and preparing the body for action. The operation of the sympathetic division is accompanied by the release of *adrenaline* into the bloodstream, which serves to maintain the activated state of the body over a longer period of time.

Synapse The term given to a junction point between two *neurones*, by means of which information is transmitted from one neurone to the next. Synapses may be *inhibitory* or *excitatory*: that is, they may render the next neurone more or less likely to fire. Normally, stimulation from several synapses (summation) will be required for the full effect on the next neurone to be achieved. [f.]

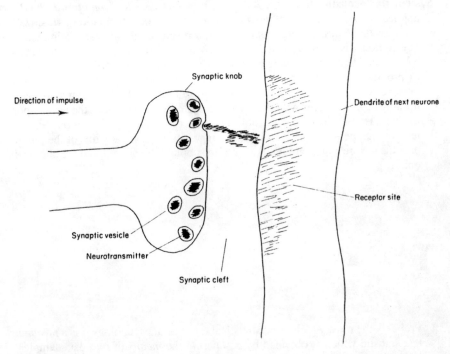

The synapse

Synaptic button See *synaptic knob.*

Synaptic cleft The small gap between the *synaptic knob* and the *receptor site* on the adjacent neurone.

Synaptic knob A swelling at the end of the *dendrite*, which contains small pockets, known as *synaptic vesicles*. Each vesicle contains a small amount of a particular *neurotransmitter*, which is released into the *synaptic cleft* when stimulated by an electrical impulse passing along the dendrite.

Synaptic transmission The transmission of information from one *neurone* to another, by means of electro-chemical processes. When the *neural impulse* reaches the *dendrites* of a given neurone, vesicles in the synaptic knob release a *neurotrans-*

mitter into the gap between it and the dendrite of the opposing neurone. Part of the dendrite is specialized to form a *receptor site*, which will pick up only that neurotransmitter, or chemicals with a similar structure. The absorption of the neurotransmitter effects chemical changes in the cell, rendering it either more likely or less likely to fire. Should enough synapses be stimulated in this way, the next neurone will either fire, or have a raised *threshold of response* such that it will not fire easily.

Syndrome A set of symptoms or events which tend to occur together functionally.

Syntax The set of rules and principles concerning the structure of a language; how the words should be combined to form what is accepted by users of the language as a grammatical sentence or phrase.

Syntactic To do with grammatical structure and organization, rather than with meaning. See *syntax*, also *semantic*.

Systematic desensitization One of the ways in which *classical conditioning* has been applied to the treatment of *phobias*. The process of systematic desensitization involves the learning (conditioning) of new responses to the feared stimulus. The new response is deliberately incompatible with the old response of fear, so that once it has been learned the phobia is extinguished. Usually, *relaxation training* is used to provide the new response, and the subject gradually learns to relax in the presence of the stimulus. A hierarchical list of feared stimuli is drawn up, and the training process begins with the least frightening situation. Once the new response to this has been learned, the subject moves on to the next situation. Since the learning happens gradually, with each stage building on the gains of the previous one, the new response gradually comes to supplant the old one and the phobia dies out.

T

t Probably the most widely used statistical test within psychology, t is a *parametric* test statistic which is obtained by comparing the *means* of two data samples, in order to determine whether any differences which occur between them are statistically *significant*. The *null hypothesis* of any given study will predict that any differences which have occurred between two sets of data have happened simply by chance. In other words, all of the scores have come from the same population, and differences between the means are simply due to random variation. On the other hand, if the means of the two sets of data are very different, it is unlikely that they have come from the same population; they are more likely to have resulted from two different populations. In that case, the null hypothesis would be refuted. The t-test looks at the mean of each set of data, bearing in mind the *standard deviation* of each one. By giving a final statistic which expresses how strong the differences between the two samples are, it allows the user to judge just how likely it is that these differences have arisen by chance. The t-test is one of the more powerful tests, in that it is able to detect significance when present; and it is also very robust (i.e. it can cope if the conditions of its use do not conform strictly to those for parametric tests).

TA See *transactional analysis*.

TAT See *Thematic Apperception test*.

T group A form of *encounter group* intended to produce a close, therapeutic relationship between the group members. T groups are traditionally free-floating and unstructured, involving a high degree of self-revelation on the part of members and with the aim of breaking down established *defence mechanisms* and barriers to open communication with other people. An alternative view, however, is that defence mechanisms should not be broken down, unless the person also receives constructive help in dealing with whatever it is that they were defending against. For this, and other, reasons T groups have been accused of being more destructive than helpful.

T maze A device used for assessing learning in laboratory rats or other animals, consisting of a straight passage from a starting box leading to a junction at which the animal is obliged to make either a right or left turn to reach a goal box, which may or may not contain a reward. [f.]

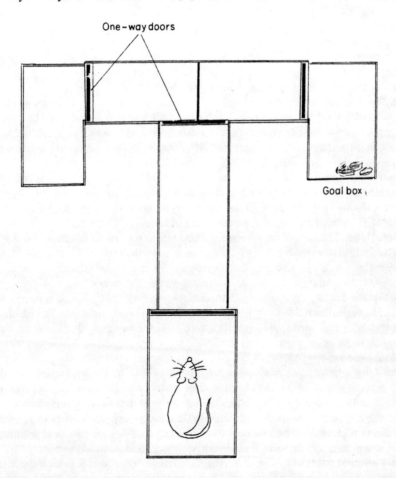

One – way doors

Goal box

A T maze

Tachistoscope (T-scope) A device used for presenting visual stimuli for precise amounts of time, and/or at precise levels of intensity. Tachistoscope studies are frequently used to demonstrate the power of *set* in perception, and were used extensively by cognitive and social psychologists during the 1950s and 1960s.

Tactile stimulation Information which is received through the sense of touch.

Telegraphic speech Concise speech which leaves out *redundant* words, as in a telegram, but gets the essential meaning across. Telegraphic speech is used by children of around 2 years, who typically combine only 2 or 3 words at a time in each utterance. Identified as such by Roger Brown, it formed the basis of his approach to language acquisition, which rejected the prevailing structural approaches to infant speech and instead focused on the child's communicative intentions. The overall approach was known as *semantic relations grammar*.

Teleology A form of logical reasoning in which the outcome is regarded as responsible for the cause, e.g. 'It rains to make the flowers grow'. This type of thinking is regarded by logicians as a mistake, by developmentalists as immature thinking, and by systems theorists as valid in some circumstances. In philosophy, teleology refers to the study of ultimate purposes; a teleological approach to scientific investigation is one in which describing the function which something serves is considered to be an adequate form of explanation, as in many biological explanations, or in *sociobiology*.

Telemetry Sending measurements over a distance, usually using radio frequencies. Telemetry is used to monitor the physiological responses of freely moving subjects such as athletes, children at play, and migrating birds.

Telepathy The communication of *cognitions* (thoughts, etc.) by other means than those understood in conventional science. There is considerable dispute over whether the phenomenon occurs, and its definition precludes rational explanation: if a communication can be explained, it is not telepathy. See also *extrasensory perception, parapsychology*.

Temperament The stable aspects of the character of an individual, which are often regarded as biologically rooted, and as providing the fundamental dispositions which, through interaction with the environment, produce the *personality*.

Temporal Concerning time, or the experience of time.

Temporal lobe The area of the *cerebrum* found below the lateral fissure at the side of each *cerebral hemisphere*. It was once thought to be the seat of the soul, and of *time perception*, although there is little evidence for this. The temporal lobe does, however, contain the *olfactory cortex* and the *auditory cortex*.

Tender-mindedness A personality characteristic put forward by William James, and later elaborated by H.J. Eysenck: characterized by a gentle, optimistic and idealistic approach to the world. Its opposite, *tough-mindedness*, is characterized by a harsher, more pessimistic approach.

Terminal button See *synaptic knob*.

Territoriality The defence and protection of one area to the exclusion of other members of the species. It has been studied mainly by *ethologists*, with particular reference to those aggressive acts which deter potential sexual competitors. The concept has been extended speculatively, and with varying degrees of sense, to account for all manner of human behaviour, ranging from international warfare to crowd violence and stress in high-density housing. See *personal space*.

Tertiary circular reactions The final stage of *circular reactions* in which the infant introduces variations in the repeated behaviour. These variations change successively, eventually becoming very different from the behaviour which the infant started out doing.

Test A standardized means of assessing the abilities or characteristics of individuals. See *intelligence, personality, psychometric, projective test, reliability, validity.*

Test administration A standard way of presenting a *psychometric* test to ensure that results obtained from subjects by different testers are comparable. See *standardized instructions.*

Test battery A combination of psychometric tests which provides a comprehensive account of an individual's functioning; such as a set of tests used for the assessment of memory disorders or reading skills.

Testee An ugly word for a person to whom a test is administered.

Test profile The portrayal (usually graphic) of the characteristics of an individual as assessed by a test or test battery. A test profile involves the presentation of a range of scores from the sub-tests rather than a single score.

Test–retest A method of assessing the *reliability* of a measure by applying the same measure, or test, on two separate occasions and *correlating* the results.

Test standardization The administration of a test to a large *sample* of the *population*, ideally a *representative sample*, which serves to provide *norms* against which the results of particular individuals or groups can be compared.

Testes Male gonads: glands which form part of the *endocrine system* of the body, and which are particularly responsible for the manufacture of *androgens*.

Testosterone A male sex hormone (androgen) which is responsible for the development of the *primary sexual characteristics* of males, and plays a large role in sexual and related activities throughout life.

Tetrachromatism A theory of colour vision which assumes that there are four primary colours: red, blue, green and yellow. See also *trichromatism.*

Texture gradient The loss of visual definition of objects with increasing distance, such that the details are seen less clearly, and general textures appear to be smoother. Gibson's ecological approach to perception argues that textural changes in the visual field are the source of most *depth perception*, and that models of perception which assume that it relies on hypothesis-testing are largely unnecessary in the explanation of experience in everyday life.

Thalamus The area of the brain found just below the *cerebrum* where information is relayed from the sensory receptors of the body to the *cerebral cortex*. It is involved in some basic information processing. Hubel and Wiesel found the basics of *pattern perception* to be related to the arrangement of *simple, complex* and *hypercomplex cells* in the *lateral geniculate nuclei* of the thalamus.

Thanatos The name of the Greek god of death, which was used by Freud to refer to the *death instinct*; a concept which he developed in order to account for the interpersonal and intrapersonal aggression of World War I.

Thematic apperception test A *projective test* in which subjects are asked to interpret and explain ambiguous scenes. The nature of their response (e.g. whether they perceive a recumbant figure as dead, drunk or sleeping) is taken as an indicator of hidden anxieties or *defences* of the *unconscious* mind. Typically, subjects will be shown about 8 or 10 different pictures, and asked to explain what is happening in each one.

Thematic qualitative analysis A form of *qualitative analysis* in which the salient material is organised into distinct themes. The themes may be data-driven, in which case they are identified during the analysis itself by grouping together recurrent ideas or concepts which seem to represent significant concerns which are being expressed by the interviewees. Alternatively, themes may be identified before the data are collected, in which case they have generally been derived from theory, and will generally relate to explicit hypotheses.

Theory An overall explanation given for a set of observations, which links them all into a coherent pattern or *model*. Scientific theories are usually considered to be of no value unless they give rise to *hypotheses* which can be tested against reality and can be shown to be false. However, exceptions to this are often made in the case of theories which are particularly appropriate to the mood of their times, such as *sociobiology*. See also *hypothetico-deductive method*.

Theory of mind A recent development of child psychology in which the child's understanding of other peoples' cognitions and emotions is the focus of study. An aspect of *cognitive development*, the child is said to develop a theory of mind between 4 and 6 years of age. The area has generated many interesting ideas and issues which are being vigorously investigated. A particular interest has been the recognition that many characteristics shown by autistic children can be summarised by the idea that they have not developed a theory of mind.

Theory of the humours A *type theory* of *personality* originating from the second century BC, and popular throughout the Middle Ages. It identified four main types of personality, each of which was supposed to come about through the action of particular body fluids. The four types are: choleric, thought to result from an excess of yellow bile; sanguine, from blood; melancholic, from black bile; and phlegmatic, from phlegm. That this was a popular theory can be seen in the way that many words have retained meanings which derive directly from that theory. It was this view of the origins of human personality which led to the word humour, which had meant bodily fluid, coming to mean 'mood' or 'temper'. [f.]

Therapeutic A term used to refer to something which is useful as an agent or tool of *therapy*.

Therapy The treatment of an individual by physical or psychological means. When applied to psychological treatments, the term implies that the client is ill and should be cured. See *behaviour therapy, client-centred therapy, cognitive therapy, family therapy, rational-emotive therapy, psychotherapy, psychoanalysis, gestalt therapy, medical model, transactional analysis*.

Thinking A general term which can be defined in several different ways: (1) the use of symbolic processes by the brain; (2) any chain or series of ideas; or (3) ideation, the sequence of producing ideas concerned with the solving of specific problems or incongruities in models of reality. Thinking is usually taken to mean conscious cognitions. Unconscious processes such as those referred to in psychoanalytic literature are seen more as *affective* responses or motivations. Most psychological investigations of thinking have concentrated on *problem-solving* or *concept-formation*. See also *creativity*.

Thought disorder A tendency to produce sequences of ideas which appear unconnected or illogical to the observer. A symptom of *schizophrenia*.

Threshold The lowest level of stimulation at which an event can be detected. Although the term 'absolute threshold' may also be used, there is nothing absolute about it. There is no fixed point at which a stimulus changes from invisible to visible, just an increasing probability that it will be detected. A threshold is therefore usually set at the point where 50% of the signals are detected by the subject. This point itself is easily influenced by factors such as sensory adaptation,

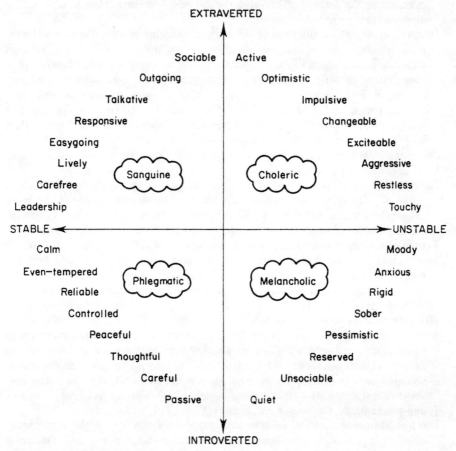

EXTRAVERTED

Sociable　Active
Outgoing　　Optimistic
Talkative　　　Impulsive
Responsive　　　Changeable
Easygoing　　　　Exciteable
Lively　　　　　　Aggressive
Carefree　Sanguine　Choleric　Restless
Leadership　　　　　Touchy

STABLE ◄――――――――――► UNSTABLE

Calm　　　　　　　Moody
Even–tempered　　　　Anxious
Reliable　Phlegmatic　Melancholic　Rigid
Controlled　　　　　Sober
Peaceful　　　　　Pessimistic
Thoughtful　　　　Reserved
Careful　　　　Unsociable
Passive　　Quiet

INTROVERTED

The theory of the humours and the EPI

set, and fatigue, so the threshold obtained will depend very much on the conditions of the experiment. See also *relative threshold, word recognition threshold.*

Timbre The tonal quality of a sound, especially used of voices. Combinations of different tones and harmonics give sounds their distinct timbres.

Time and motion A method of analysing working patterns developed by F.W. Taylor (1903), involving a systematic breakdown of movements and skills. By breaking down work sequences into sequences of actions with maximum economy and minimum effort, Taylor showed how productivity in industry could be dramatically increased, and his work is often considered to be the foundation of *ergonomics.* Although it is still used from time to time today, it has proved to be of only limited value on the factory floor; people have an understandable aversion to being treated as if they were robots.

Time perception The subjective awareness of the passage of time, which is found not to correspond precisely with objective time. Time perception has been studied experimentally to determine the effect of various forms of *cognitive* tasks, and of *psychoactive drugs.*

Tip of the tongue phenomenon A phenomenon of memory in which the individual experiences the feeling of knowing the desired information but is temporarily unable to bring it to consciousness.

Token economy The application of B.F. Skinner's techniques of *operant conditioning* to establish an environment in which desired behaviour is *reinforced* with tokens which can then be exchanged for goods or privileges. Token economy is proposed as an efficient way of modifying behaviour, especially of long-stay *institutionalized* patients. There is some evidence, though, that any improvements found in patients using token-economy systems may be a by-product of raising staff morale through providing them with apparently meaningful tasks, rather than through the direct operation of conditioning.

Tolerance Adaptation to the effects of a drug so that increasing doses are needed to achieve the same effect.

Tomography A method of investigation of brain functioning, mostly used for medical purposes, which involves building up a three-dimensional picture of the brain through a succession of X-ray photographs or ultrasound images, in order to identify abnormal structures or growths.

Tone The quality of sound in terms of the number of different frequencies which make up that sound. A pure tone consists of sounds of one frequency only, but most sounds are combinations of several different frequencies at differing strengths, often with one frequency being dominant over the others, which becomes identified as the pitch of the sound.

TOT See *tip of the tongue phenomenon*.

TOTE Abbreviation of 'Test Operate Test Exit', proposed as the basic component of planned actions by Miller, Galanter and Pribram in *Plans and the Structure of Human Behaviour*. They proposed that the organism Tests the environment; Operates on it to bring about change; Tests again to see whether the outcome is satisfactory; and, if so, Exits from the sequence. See *negative feedback*.

Tough-mindedness See *tender-mindedness*.

Trace conditioning A form of classical conditioning in which the *conditioned stimulus* is presented immediately before the *unconditioned stimulus*, rather than simultaneously. See *delayed conditioning, simultaneous conditioning*.

Trait An aspect of *personality*, such as sociability, impulsiveness, conventionality, etc. See also *trait theory*.

Trait theory A theory of *personality* in which personality is considered to consist of a collection of differing, usually measurable traits. One of the best known examples, is that of R.B. Cattell whose *personality inventory* measures 16 different personality factors (and so is called the 16PF).

Trance An altered *state of consciousness* in which *decision-making* and executive functions are partially suspended, and *attention* is highly focused. Trance can be achieved by *hypnosis*, meditation, some drugs, and some clinical conditions.

Tranquillizer A drug used to reduce stress or anxiety temporarily; usually a *benzodiazepine* such as Valium or Librium.

Transaction A chain or sequence of interactions between two or more individuals, which is based on the idea that *interactions* can only be understood if it is recognized that each person influences the other. The behaviour to which a person responds is itself a result of an earlier influence. For example, a baby becomes upset while feeding, the mother responds by being anxious during feeds, this makes her treat the baby differently, causing further upset to the baby and so on. In general, transactions mean that two people negotiate the environment in which they will both have to function in the future, rather than one person defining the environment for the other. The concept of transactions is particularly useful in the analysis of interactions between parent and infant; but can be applied to most areas of social behaviour. See *coevolution*.

Transactional analysis A scheme developed by Eric Berne for interpreting interpersonal interaction. Each person in a given pair, or dyad, may operate as a child, an adult, or a parent. Acting in a submissive, dependent manner (child) may provoke the partner of the dyad to adopt a parental role; acting in a dominating manner (parent) may produce submissive behaviour from the other, etc. This model has been usefully applied in individual and group *psychotherapy*, particularly by uncovering the recurrent patterns of social interactions described in *Games People Play*.

Transactional theory A theory of perception developed by Ames which states that perception develops as a result of constant interaction with the environment.

Transcendental meditation A technique derived from Hindu practice, of sustained concentration on a brief phrase (*mantra*) to induce *relaxation*. See *states of consciousness, hypnosis, trance*.

Transducer A device to convert a biological signal such as heartbeat or skin resistance into an electrical signal suitable for recording. More generally, any sort of device which converts energy into a different form.

Transduction A term which is usually used of sensory receptors, e.g. in referring to the conversion of photons (light) into electrical energy by rod and cone cells in the retina. See *transducer*.

Transfer of training A phenomenon in which the learning of one particular task either helps or hinders the learning of a subsequent task. Positive transfer involves the facilitation of subsequent learning, while negative transfer impedes it. See also *proactive interference*.

Transference The way in which feelings derived from a previous relationship may be transferred to someone new. This is particularly used in *psychoanalysis*, in which the analyst deliberately maintains a neutral, colourless personality so that such transfer becomes easy for the patient, e.g. when the therapist is responded to as the patient's father. Transference is similar to *projection*, and was first regarded by Freud as a nuisance, but is now regarded as an essential source of information about the patient's early relationships. Interpretation of the transference has been claimed as the major or only source of therapeutic change. See *counter-transference*.

Transformational grammar A set of rules which specify how one sentence can be transformed into another in a language, particularly converting *deep structure* into *surface structure*. The concept was originally put forward by Chomsky as part of his explanation of how children acquire language. A complete set of transformational rules amounts to a grammatical description of a language.

Transgenerational transmission The passing on of environmentally acquired characteristics to subsequent generations, e.g. the underfeeding of one generation of rats at a *critical period* may result in reduced size of the adult rats two generations later; and extreme malnutrition for a pregnant woman may have subsequent effects on the child.

Transmitter substance See *neurotransmitter*.

Transsexual A person who changes sex, either from male to female or from female to male, usually through a course of hormone therapy and surgery. Although typically transsexuals have always experienced themselves as being 'really' the other sex, the main part of transsexualism involves the learning of a new sex role. Many transsexuals have to spend an extensive period of time, at least a couple of years, passing as a member of their desired sex before being allowed treatment.

Transvestite Transvestites are people who enjoy dressing as members of the opposite sex, and may do so quite elaborately. Most transvestites are *heterosexual*, although transvestism can sometimes be associated with *homosexuality*. Transvestites in general tend to be contented with their own sex and sex role and do not usually experience problems of sexual identity.

Trauma An experience which, because of its intensity and unexpectedness, is damaging. The initial reaction is shock, which may or may not be followed by recovery. Freud came to believe that all *neuroses* were caused by childhood traumas which remained unresolved in the adult. In medicine, the term means bodily injury caused by an external object.

Trial and error learning Learning which takes place as a result of trying out a variety of responses to a given stimulus, until one response achieves the desired effect, whereupon it becomes more likely to be repeated. Thorndike proposed that trial and error was the basis of all learning, but work on *latent learning* by Tolman brought this into question.

Triarchic intelligence A theory of *intelligence* outlined by Robert Sternberg (1985), which consists of three separate subtheories. Each subtheory concerns a different aspect of manifest intelligence: (1) contextual intelligence, concerned with intelligence in its sociocultural setting; (2) experiential intelligence, which is concerned with how the individual's own past experience influences how they approach a given task or situation; and (3) componential intelligence, concerned with the cognitive mechanisms by which intelligent behaviour is achieved. The componential subtheory incorporates an earlier theory of intelligence (Sternberg, 1977), in which components of intelligence are classified in terms of function and level of generality, resulting in three categories of components: metacomponents, or higher-order processes involved in, for instance, planning and decision-making; performance components, those involved in actually carrying out a task; and knowledge-acquisition components, involving the processes involved in learning new information.

The triarchic theory is therefore distinctive, in that it treats intelligence as mental activity which is directed towards purposive activity in the real world, rather than as a reified, context-free cognitive exercise. By integrating sociocultural and experiential intelligence with the specific tasks generally involved in intelligence testing, it also provides a theoretical framework for the selection of appropriate content for *intelligence tests*. See also *reification*.

Trichromatism A theory of *colour vision* which proposes that it results from perceiving combinations of just three colours: red, blue and green (not the same as the primary colours of red, blue and yellow for pigments).

Turner's syndrome A genetic disorder in which the individual has one fewer *chromosome* than normal, resulting in sexual abnormalities.

Two-factor theory A model of intelligence put forward by Spearman, who argued that any particular intelligent act originates from two different intelligence factors: a 'g' (general) factor, common to all behaviour, which is characteristic of the individual's general functioning, and an 's' factor, specific to the problem in hand, which is the relevant skill for that particular behaviour, e.g. mathematical, verbal, spatial, manipulative, etc.

Two-process theory of memory First proposed by James, and developed further particularly by Miller and Atkinson and Shiffrin, this theory holds that two distinct forms of memory exist, each with its own characteristics: immediate, or *short-term memory* (STM); and *long-term memory* (LTM). There is dispute as to how far these forms of memory are in fact distinct. See also *levels of processing*.

Two-tailed hypothesis An hypothesis which predicts a result from either end of a frequency distribution of the *null hypothesis*, e.g. a prediction that scores will vary on a task from one day to the next would be two-tailed; whereas a prediction that scores would improve from one day to the next would be one-tailed, as only one of the two kinds of outcomes (improving or getting worse) is predicted.

Type A and type B behaviour As outlined by Friedman and Rosenman, these refer to styles of working shown by company executives. Type A individuals are typically anxious, driving people, who find it difficult to delegate tasks to other people, and tend to worry about their work when at home. Type B individuals may work just as hard, but have a more relaxed style, and an easygoing approach to problems, dealing with each as it arises rather than worrying about them all. Friedman and Rosenman found these styles to correlate strongly with susceptibility to heart disease: Type A individuals being far more likely to suffer heart attacks than Type B individuals.

Type 1 error A statistical term for the mistake of rejecting the *null hypothesis* when it should have been retained. In *experimental* as opposed to *correlational* studies, this would mean concluding that a difference in the *dependent variable* is attributable to the *independent variable* when in fact it was due to other factors, e.g. deciding that a difference in the performance of two classes was due to different teaching methods when in fact it derived from *individual differences* in the students concerned. *Significance levels* are usually set so as to make Type 1 errors unlikely, but in some circumstances you would tolerate a high risk of Type 1 error (e.g. a cheap, safe and easy way to prevent cancer). See *statistical significance*.

Type 2 error A statistical term for the mistake of retaining the *null hypothesis* when it should have been rejected. In experimental studies, this would mean concluding that the *independent variable* had no effect on the *dependent variable*, when in fact it did have some influence. This is usually regarded as the less costly kind of error (see *Type 1 error*). Note that this is not a statistical error, as statistics merely assess the probability of relationships; the error arises in the conclusions that are drawn from the statistics.

Type theory A theory of *personality* in which people are classified according to common characteristics. Sheldon grouped people according to types of physique, their *somatotype*, with personality characteristics supposedly associated with particular kinds of bodily build. Jung also grouped people according to personality type, most famously *introversion* and *extraversion*. The *theory of the humours* provides another example of an early type theory of personality. A more restricted approach in the study of personality is the 'narrow-band' approach, the identification of a single type such as the *authoritarian personality*.

U

Unconditional positive regard A prerequisite for mental health and personal growth, according to Carl Rogers. Rogers identifies two basic human needs: the *need for positive regard*, from other people, and the need for *self-actualization*. The person must satisfy both of these needs, but if their only positive regard is *conditional* upon 'good' or appropriate behaviour, then much of their behaviour will be directed towards obtaining that approval. This means that they will not feel free to explore their own potential and their need for self-actualization because of the fear of engendering social disapproval. Most people, however, have at least one person at some time in their life who gives them unconditional positive regard. In that relationship, they can be sure of the other person's affection and warmth, and this means that they can feel free to develop and explore new aspects of themselves. Unconditional positive regard is usually provided by parents, during childhood, though Rogers believes that it is not tied to the early years of life. The formation of such a basis of unconditional positive regard is at the heart of Rogers's *client-centred therapy*.

Unconditioned response A response which occurs automatically to a particular stimulus, and does not have to be learned. For example, pulling the hand away from a burningly hot surface is an unconditioned response: it happens as a *reflex*, without the need for conscious recognition of what is happening. See also *unconditioned stimulus, conditioned response, conditioned stimulus, classical conditioning*.

Unconditioned stimulus A *stimulus* which automatically produces a response in an organism (animal or human being). The term 'unconditioned' means 'not learned': a stimulus of this kind will operate by reflex, or automatically, with no learning being necessary. It forms the basis for *classical conditioning* as the new, *conditioned* stimulus becomes linked with the unconditioned one.

Unconscious A lack of *conscious* awareness. The most important use of the term is in psychoanalytic theory as a reference to mental activity which is not available to consciousness because it concerns material which is too threatening to the *ego* to be recognized directly. Freud believed that the unconscious has its own way of working (see *primary process*) which is different from that of the conscious mind. For example there is no awareness of time in the unconscious so all threats are felt as if they were still present, even if the source of the threat disappeared years ago. See also *preconscious, psychoanalytic theory*.

Unconscious motive A motive of which the person is unaware but which continues to have an effect on behaviour. For example, a student may under-achieve during exams owing to an unconscious rebellion against parental pressure to succeed; although consciously, she or he will be trying to do as well as possible, the chosen revision strategies are ineffectual, relying on rote learning or simply reading through notes, and this ensures that they do not do as well as they could. Unconsciously, they have shied away from being too successful. Human behaviour is often influenced by such unconscious motives; and disentangling them such that the individual becomes aware of what is going on can be one of the main tasks of a *psychotherapist*.

Unipolar depression A depression which is similar in form to a *bipolar disorder* but in which the manic phase is absent; the person simply has the depressive periods without the swings to mania.

Utterance Something which is said; a simple unit of speech or language. The term is often used when describing spoken language, as it avoids making assumptions about the grammatical form of what was said. Describing something as a 'sentence' or a 'phrase' might not be accurate, but calling it an 'utterance' merely makes the assumption that it was actually uttered.

V

Validity Validity refers to how far a given measure assesses what it was intended to measure. There are generally considered to be three main types of validity: surface or face validity, criterion validity, and construct validity. Surface validity, is judged simply in terms of how far the measure seems appropriate: it is an assessment of how plausible the chosen measure is. A questionnaire item asking how people feel about sex discrimination has surface validity in that it appears, on the surface, as though it will allow us to find out about sex discrimination.

Criterion validity is when the measure being used is compared with some other measure or standard which assesses the same thing. Criterion validity may be of two basic forms. Predictive validity involves the measure being compared with some future event, such as assessing the validity of IQ tests by looking at how well they correlate with later examination success. Concurrent validity involves the measure being compared with a measure obtained at the same time, such as comparing stated attitudes towards sex discrimination with behavioural measures of participation in housework. More simply, a new test may be compared with the results from an existing and widely accepted test.

Construct validity is how far the measure being examined truly represents the theoretical construct which it is supposed to measure. The most well-known example of this is the question of how far intelligence tests truly measure intelligence. In this case, 'intelligence' would be the theoretical construct, and the IQ score the measure being assessed for construct validity. The assessment would be made by examining whether people with higher IQ scores in fact behave in ways that would be judged as more intelligent.

Variable Anything which varies; something which can have different values. Any measure of performance or behaviour taken in a study is referred to as a variable, because it can have different values depending on circumstance. If its value depends on the particular experimental situation which was set up, then it is known as the *dependent variable*. The conditions set up by an experimenter in a formal experiment also vary: typically there is an experimental and a control condition, and often there may be several variations of the experimental condition. For this reason that, too, is known as a variable, the *independent variable*. Other features of a study can vary too, like background noise or time of day. If these are randomly distributed, so that they can affect any of the conditions of the independent variable equally, then they are referred to as random variables. But if they are likely to affect certain conditions of the independent variable more than others, they are known as *confounding variables*.

Variable-interval reinforcement A *reinforcement schedule* in which the delivery of a *reinforcement* depends on the amount of time which has passed since the last one was given. The amount of time between each rewarded response varies, but works out to a set average within a given time period. For instance, a VI10 schedule would mean that an average of 10 seconds would have to pass after each reinforcement before another reward could be obtained; but the actual time might be less or more than that on any given trial. Variable-interval reinforcement schedules tend to produce a steady *rate of response*, which is highly *resistant to extinction*. See also *partial reinforcement*.

Variable-ratio reinforcement *Reinforcement* given during *operant conditioning* in such a way that not every response made is reinforced, but only a certain proportion of them. The ratio of reinforced to non-reinforced responses varies after each reinforcement, and will average out to a pre-set proportion. For instance, a *reinforcement schedule* of VR10 would mean that, on average, one response in every 10 would be rewarded; but the number of responses which had to be made before each reward given was randomly varied around ten. Variable-ratio reinforcement produces a very rapid *response rate* which is highly *resistant to extinction*. Many naturally occurring reinforcement schedules are in the form of a variable ratio. For example a child who demands attention may only receive it unpredictably. A high level of demandingness, which is very resistant to extinction would then be predicted. See also *partial reinforcement*.

Variance A *parametric* measure of *dispersion*, obtained by subtracting each score from the *mean* of the sample, squaring these differences, and averaging them. Mathematically it has the advantage that scores that are only slightly different from the mean have very little effect, while more extreme scores (because the differences are squared) have more influence. The square root of the variance is the *standard deviation*.

Ventromedial nucleus A region of the hypothalamus which seems to be particularly concerned with *satiation*: stimulation of this region in rats will cause them to cease eating even when previously they have shown strong signs of hunger and lesions in the region result in rats becoming obese through overeating. It is thought that some cases of human obesity may arise from some kind of disorder within this region of the *hypothalamus*.

Verbal deprivation hypothesis The idea put forward by Bernstein and others, that the form of language learned by a child could represent a disadvantage when it came to learning or handling abstract forms of information. Bernstein argued that *restricted codes* of language, with their relatively limited vocabularies and reliance on shared assumptions on the part of the listener, meant that children would find the kind of conceptual and abstract learning which they encountered in school inherently more difficult than children who use *elaborated codes*. This idea was hotly disputed by many researchers, notably Labov, who demonstrated that children who used highly restricted codes of language, such as Black American English, were perfectly capable of handling abstract and theoretical concepts, as long as those concepts were introduced in a setting in which the children felt confident.

Verbal memory This term is used in two main ways. Firstly to mean the storing of mental images by using words as a form of coding for information. In this case, verbal memory is simply meant as a variation of *symbolic representation*, with all the associated features and advantages. Secondly, the term is used to mean memory for words. Much laboratory research on memory, especially in the early years, concentrated initially on asking subjects to memorize lists of words, partly because subjects were able to state clearly exactly what they remembered, which was not always easy with visual or auditory images. But there is considerable recent evidence to suggest that this form of learning is qualitatively different from the way that people remember connected prose or speech, and even more different from everyday memory. In its most general sense, verbal memory includes memory for speech and prose.

Vicarious learning Learning through observing what happens to others. Vicarious learning was particularly investigated by Bandura in studies of imitation in children. He found that children who saw others being rewarded for aggressive acts were more likely to *imitate* them. Behaviour patterns may be acquired or abandoned as a consequence of seeing other people being rewarded or punished for them. See also *identification*.

Vigilance See *sustained attention*.

Visual cliff Apparatus designed by Gibson and Walk to investigate whether animals have an *innate* perception of depth. A newborn animal (e.g. a chick or goat kid) is placed on a centre board over a sheet of strong glass which covers a steep drop. If the animal shows fear or refuses to cross over the drop it is assumed that the ability to perceive depth is present. Since the animal is newly born, this cannot have been learned, and therefore must be regarded as innate. Results are more difficult to interpret when human babies are used since they are not mobile at birth. [f.]

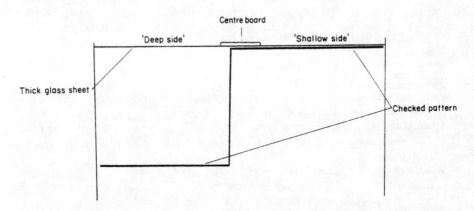

A cross section of the 'visual cliff'

Visual cortex The part of the cerebral cortex which is responsible for the decoding of visual information. The visual cortex is found at the back of the brain, in the *occipital lobe*. Also known as the striate cortex, this area forms the main *sensory projection area* for vision and *electrical stimulation* of this area can often produce vivid visual sensations.

Visual field The 'scene' or expanse of visual information which is encompassed by the *retina* at any moment. When we are looking at something, the object of our attention is at the centre of the visual field, and we see it most clearly. But we also receive a visual impression of the surroundings, and this stretches for quite a long way around the focus of vision. A slight movement at the side of the visual field will usually cause us to turn slightly and focus on a new centre of visual attention; the visual field then covers a different, but overlapping, range of visual stimuli.

Visual illusions Figures which appear to be other than they really are, as a result of the ways in which the brain interprets information. Visual illusions have been extensively studied by psychologists, as it is thought that investigation of the errors of perception can throw light on how normal perceptual processes work. The visual illusions most commonly studied by psychologists fall into three main categories: geometric illusions, usually in the form of simple line drawings, illusions of movement, such as the *phi phenomenon* or the *waterfall effect*, and colour illusions. [f.]

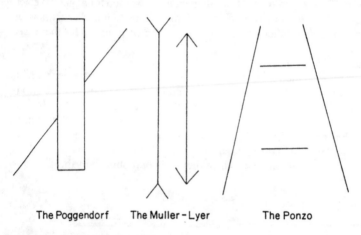

The Poggendorf   The Muller–Lyer   The Ponzo

Some well-known visual illusions

Visual perception The analysis and interpretation of information received and processed through the *visual system*. See also *perception*.

Visual stimulation Any form of light which reaches the *retina* and causes the *rod* or *cone cells* to react. The term is usually used to refer to a visual image which is received by the eye. See also *stimulus*.

Visual system The general name given to the set of *neurones* and brain structures involved in the processing of visual information. The visual system includes the eye, in particular the *retina*, the *optic nerve*, the *optic chiasma*, the *lateral geniculate nuclei* of the *thalamus*, and the *visual cortex*. [f.]

VMH See *ventromedial nucleus*.

Vocalization The production or articulation of audible speech sounds. The term is particularly used when referring to the *babbling* or crying noises made by babies before they have recognizable speech.

Voice-recognition systems Computer systems which can analyse the distinctive features of the human voice, and respond to key words which have been spoken. The development of voice-recognition systems forms a major area of research in the field of *artificial intelligence*, but represents no easy task, owing to wide differences in articulation shown by different people. Some success has been

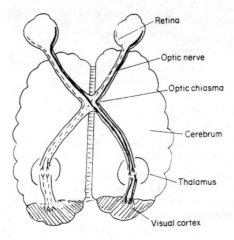

Retina

Optic nerve

Optic chiasma

Cerebrum

Thalamus

Visual cortex

The visual pathways from the eyes to the visual cortex

achieved in the development of systems which can 'learn' patterns of speech used by a particular person. This is usually achieved by the individual concerned reading out a set of key words and phrases, which the computer system uses as a baseline for identifying their characteristic speech patterns, and retains for when next receiving information from that person. See also *expert systems, artificial intelligence*.

Voice stress analyser A device for analysing the acoustic properties of the voice, which examines in particular the minute variations of tone and pitch occurring as vowel sounds are produced. In a relaxed voice, there are many of these variations, but they occur rapidly and the overall impression is that the sound is smooth and regular. The variations can be seen clearly on a spectrograph, which gives a visual image of the sound. If someone is under stress, even though they may try to keep their voice sounding normal and although it may sound the same to a listener, the effort of keeping their vocal cords steady will mean that, when analysed by a spectrograph, the sound appears to be 'flattened out', without the normal small variations. Accordingly, the analysis of speech by a spectrograph provides a sensitive and reliable measure of stress. See also *GSR, polygraph*.

Voice timbre See *timbre*.

Volley principle In the *auditory system*, and in several other sensory systems, the intensity of a *stimulus* is signalled by means of the rate at which *electrical impulses* are fired to the brain. This signal can be achieved by each *neurone* firing very rapidly but owing to the *absolute* and *relative refractory periods*, there is a limit to how fast each neurone can fire. In the case of very intense stimuli, the neurones fire in relays or volleys: a set of neurones will fire, closely followed by another set, and then other. In this way, the brain receives a series of impulses at a rate which would not be possible for the neurones if each were firing singly.

Voluntary behaviour Behaviour which forms a deliberate action on the part of the individual. Such behaviour is usually contrasted with *involuntary*, or *reflexive*, behaviour. *Operant conditioning* and the higher forms of learning are usually concerned with the training of voluntary behaviour except in the case of *biofeedback; classical conditioning* is typically concerned with involuntary behaviour.

WAIS See *Wechsler adult intelligence scale*.

Waterfall effect A special case of a *negative aftereffect* involving the perception of steady movement. If someone looks steadily at movement which occurs consistently in one direction, such as gazing at a waterfall, then, when they look away at a stable background, they experience an illusion of movement in the opposite direction. In the case of the waterfall, this involves the impression that the bank or surroundings are moving steadily upwards; if the effect is as a result of the movement of a train, then the train may seem to be moving backwards when it stops at a station.

Weber's law A law discovered by Ernst Weber in the early years of psychology, during which *psychophysics* was being developed. The law states that the amount by which a stimulus needs to be changed in order for the change to be noticeable (the *just noticeable difference*) is a constant proportion of the strength of the stimulus. The value of this constant proportion is known as Weber's constant. In practical terms, the implications of Weber's law is that stronger stimuli will need to increase or reduce by greater amounts than do smaller stimuli, before they are perceived as different. See also *Fechner's law; power law*.

Wechsler adult intelligence scale (WAIS) One of the major *intelligence tests*, produced by David Wechsler. Although it produces an overall IQ score, this can be sub-divided under two general headings of verbal IQ and performance IQ, each of which is composed of different sets of items (6 for verbal and 5 for performance). In principle it is possible to identify specific kinds of disability or deficit using such tests; but in practice, for differences between subset scores to be big enough to be significant, the deficit in the person's performance would be obvious anyway. [f.]

Wechsler intelligence scale for children (WISC) A version of the WAIS designed for use with children. It will measure IQ from 6 to 16 years.

WEG Acronym for warmth, empathy and genuineness, the three therapist attributes which have been proposed as the most important factors in the effectiveness of *psychotherapy*. WEG is thought to be more important than any specific therapeutic technique.

Wernicke's area The area of the *cerebral cortex* which is particularly concerned with the interpretation and understanding of language. Damage to this area produces *aphasia* or difficulties in the comphrehension of speech. See also *Broca's area, angular gyrus*.

White matter The term used to refer to the densely packed masses of *myelinated* nerve fibres which are found in the *central nervous system*. In the *brain*, this is found on the inside, with grey matter (consisting of unmyelinated fibres and cell bodies) covering the outer surface. In the *spinal cord* this is reversed: the white matter being on the outside and the grey matter being to the inside, surrounding the central canal.

WISC See *Wechsler intelligence scale for children*.

Withdrawal symptoms Temporary physical disorders which occur as a result of the subject failing to receive their normal dose of a drug on which they have become *dependent*. Withdrawal symptoms can be quite severe, depending both on the drug

Test	Scaled score	
Information		
Comprehension		
Arithmetic		
Similarities		
Digit span		
Vocabulary		
Verbal score		
Digit symbol		
Picture completion		
Block design		
Picture arrangement		
Object assembly		
Performance score		
Total score		

Verbal score _____ IQ _____

Performance sore _____ IQ _____

Full scale score _____ IQ _____

A test record sheet for the Wechsler adult Intelligence Test

concerned and on the extent to which the subject has become *habituated* to the drug. Many ordinary drugs, such as caffeine, can produce quite strong withdrawal symptoms if the individual has previously had a high regular intake and suddenly ceases to take the drug altogether. The existence or otherwise of withdrawal symptoms is one of the main indicators of physiological *addiction*.

Wolf children Children found living in the wild, whose behaviour led people to believe that they had been brought up by wolves. Such children would be of great interest to psychologists to test critical periods for abilities such as language acquisition. However it is suspected that in most cases the child was abandoned quite recently because of severe mental disturbance, and that it is the disturbance which is responsible for the unusual behaviour and restricted abilities.

Word recognition threshold A measure of the minimum degree of exposure of a word necessary for a subject to identify it. The normal procedure is to vary the time during which the word is exposed; other conditions could involve presenting the word more or less faintly, or at different distances. The *threshold* is usually taken to be the point at which the word is recognized 50% of the time, as the exposure necessary will vary according to the conditions under which the word is presented. Recognition thresholds are usually measured using a *tachistoscope*.

X A term normally used to represent a raw score in a set of data, usually plotted as the *abscissa* (or horizontal axis) of a graph; also used for any unknown score or the value of an *independent variable*.

x̄ An abbreviation often used to refer to the *mean* of a set of scores.

X chromosome A distinctive chromosome named for its appearance under the microscope, which carries information directing the development of sexual characteristics. In women, the X chromosome is paired with another, similarly-structured X chromosome, but in men it is paired with a small, truncated chromosome known as a *Y chromosome*. [f.]

Xenophobia An irrational and excessive fear of strangers or strange, (foreign) cultures, which can often become converted into intense, jingoistic patriotism, and/or racial or cultural *prejudice*.

XX An abbreviated reference to the combination of chromosomes shown by women: men are referred to as XY. See also *X chromosome, Y chromosome*.

XXY syndrome See *Klinefelter's syndrome*.

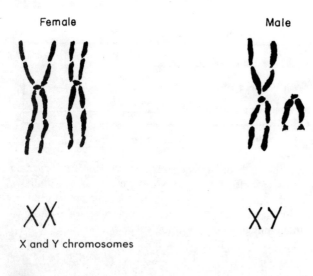

X and Y chromosomes

Y A term used for scores from a second set of data, usually plotted as the ordinate (or vertical axis) of a graph; x is the term used to refer to those from the first set.

Y chromosome A distinctively-shaped chromosome whose presence as one of a pair indicates that an individual is male. The other one of the pair will be an *X chromosome*. See also *sex-linked trait*.

Yavis A term used to describe the typical patient considered suitable for *psycho-analysis*. The term stands for 'young, attractive, verbal, intelligent and successful'. Patients who don't fit these criteria are frequently allocated to other less expensive forms of treatment, e.g. *behaviour therapy*. Although this idea is only semiserious, it contains more than a grain of truth in terms of the types of patients which many psychoanalysts feel they can be most effective with.

Yerkes–Dodson law An expression of the relationship between a person's state of physiological *arousal*, and their performance of a task or job. When plotted on a graph, it takes the form of an inverted-U curve. Up to a point, increased arousal improves performance, but beyond that point further increases in arousal will cause performance to deteriorate. Furthermore, the shape of the curve will vary with the complexity of the task, simple tasks being less affected by high arousal, and showing a wider flatter curve and complex tasks reaching their optimal level at a relatively lower state of arousal, increasing and falling off more sharply. [f.]

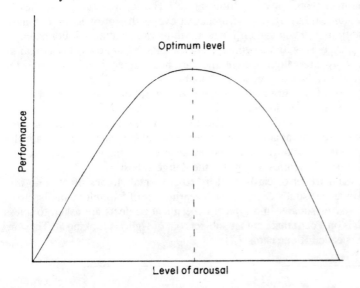

The Yerkes–Dodson Law

Yerkish An artificial 'language' developed during experiments in chimpanzee language training at the Yerkes primate laboratory in Georgia, USA. Initially developed for use with a chimpanzee named Lana, it consists of a series of symbols used in a fairly arbitrary fashion to stand for concepts and conjunctions. There is considerable debate as to just how far Yerkish can be regarded as a language rather than just an arbitrary set of symbols.

Yoked control An experimental set-up in which the experimental group and the control group are paired, such that any member of the experimental group has one of the control group receiving exactly the same experiences. The pairs are linked ('yoked') in such a way that what happens to one also happens to the other: e.g. if one receives a reward or punishment the other does too. This makes it easier for the experimenter to ensure that any differences which arise between the two are produced by the *independent variable*, rather than by variations in experience.

Young–Helmholtz theory A theory of colour vision which argues that colour is perceived through the stimulation of receptors which are sensitive to red, green and blue light. Other colours can be perceived by combinations of these three, in the same way as the coloured dots on the screen of a colour television produce a complete spectrum by combination. See also *opponent-process theory*.

Z

Zeitgeber A German word meaning 'time giver'. The term occurs in studies of *biological rhythms*, referring to environmental events that provide the organism with a precise timing to which their innate rhythms can be attached. For example many animals are born with a rhythm of approximately 24 hours (the *circadian rhythm*). The daily alternation of light and dark is a zeitgeber which enables the circadian rhythm to adjust to precisely 24 hours.

Zeitgeist The 'spirit of the times'; the social and cultural climate within which an event occurs or a theory is developed. Scientific theories are very rarely, if ever, independent of their cultural climate and the form that a theory takes and the information which counts as acceptable evidence for a theory can vary dramatically from one period to the next. By and large, those scientific theories which become popular tend to be the theories which 'fit' the Zeitgeist best.

Zener cards A standard set of cards used in experimental studies of *extrasensory perception*. There are usually 25 cards, each bearing one of 5 simple symbols: cross, wave, circle, star or square. In a typical experiment subjects are asked to guess which pattern is on a card that another subject (out of sight) is looking at. They are also sometimes called Rhine cards. [f.]

Zener cards

Zero-sum game In *games* theory, the class of games in which a fixed quantity of resources is distributed between the players, so that for anyone to do better, someone else must do worse. Zero-sum games are of particular interest to social psychologists because it has been found that people operate according to the same principles even when they are not in a zero-sum situation. That is, people will often work hard to ensure that others do worse than themselves even if this has no effect on their own gains and in some cases may even mean sacrificing them.

Zöllner illusion A particularly powerful visual illusion in which parallel lines appear to converge as a result of being crossed by short diagonal lines set at angles to the main ones. [f.]

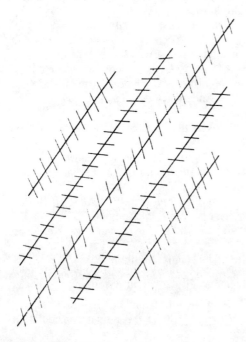

The Zöllner illusion

Zoology The study of animals. Zoology includes the study of animal physiology and animal behaviour. In the latter area it frequently overlaps with *comparative psychology*.

Zurich school The group of psycholanalysts who joined Carl Gustav Jung as he developed an alternative school of analysis after his break with Freud. Jung's method is referred to as *analytical psychology*.

Zygote An *ovum*, or egg, which has been fertilized and so is capable of developing into a young member of the species.

BIBLIOGRAPHY

Atkinson, R.C. and Shriffin, R.M. (1971). The control of short-term memory. *Scientific American*, 224:81–90.

Bandura, A. (1977). *Social learning theory*. Englewood Cliffs, NJ: Prentice-Hall.

Bartlett, F.C. (1932). *Remembering: A study in experimental and social psychology*. Cambridge: Cambridge University Press.

Beck, A.T. (1976). *Cognitive therapy and the emotional disorders*. New York: International Universities Press.

Berkowitz, L. (1965). The concept of aggressive drive. In Berkowitz. L. (ed.), *Advances in experimental social psychology, Vol. 2*. New York: Academic Press.

Bernstein, B. (1961). Social class and linguistic development: a theory of social learning. In A.H. Halsey, J. Floud and A. Anderson (eds), *Education, economy and society*. Glencoe: Free Press of Glencoe.

Binet, A. and Simon, T. (1905). New methods for the diagnosis of the intellectual level of subnormals. *Annals of Psychology*, 11:191.

Bowlby, J. (1953) *Child care and the growth of love*. London: Penguin.

Broadbent, D.E. (1958). *Perception and communication*. London: Pergamon Press.

Bruner, J.S. (1973). *Beyond the information given*. Toronto: Mcleod.

Burt, C. (1958). The inheritance of mental abilities. *American Psychologist*, 13:1–15.

Campbell, H.J. (1973). *The pleasure areas*. London: Eyre Methuen.

Cannon, W.B. (1929). *Bodily changes in pain, hunger, fear, and rage*. New York: Appleton.

Chomsky, N. (1957). *Syntactic structures*. The Hague: Moulton.

Cumming, E. and Henry, W.H. (1961). *Growing old: The process of disengagement*. New York: Basic Books.

de Bono, E. (1966) *Lateral thinking*. Harmondsworth: Penguin.

Deutsch, J.A. and Deutsch D. (1963) Attention: some theoretical considerations. *Psychological Review* 70:80–90

Donaldson, M. (1963). *A study of children's thinking*. London: Tavistock.

Eckman, P. (1982). *Emotion in the human face* (2nd ed.). New York: Cambridge University Press.

Erikson, E.H. (1963). *Childhood and society* (2nd ed.). New York: Norton.

Eysenck, H.J. (1953). *The structure of human personality*. London: Methuen.

Festinger, L. (1957). *A theory of cognitive dissonance*. Stanford: Stanford University Press.

Gesell, A. and Ilg, F.L. (1949). *Child development*. New York: Harper.

Gregory, R. (1966). *Eye and brain: The psychology of seeing*. Weidenfeld & Nicolson.

Hebb, D.O. (1949). *The organization of behaviour*. New York: Wiley.

Jensen, A.R. (1969). How much can we boost IQ and scholastic achievement? *Harvard Educational Review*, 39:1–123.

Kelly, G.A. (1955). *The psychology of personal constructs*. New York: Norton.

Kempe, C.H. *et al.* (1962). The battered child syndrome. *Journal of the American Medical Association*, 181:71.

Kohler, W. (1925). *The mentality of apes*. New York: Harcourt Brace. (Reprint ed., 1976. New York: Liveright).

Labov, W. (1972). *Language in the inner city: Studies in the black English vernacular.* Philadelphia, PA: University of Pennsylvania Press.

Laing, R.D. (1960). *The divided self.* London: Tavistock.

Lashley, K. (1929). *Brain mechanisms and intelligence: A quantitative study of injuries to the brain.* Chicago, IL: University of Chicago Press.

Lorenz, K. (1937). *On Aggression.* New York: Harcourt Brace.

McGarrigle, J. and Donaldson, M. (1974). Conservation accidents. *Cognition,* 3: 341–50.

Maslow, A. (1962). *Toward a psychology of being.* New York: Van Nostrand.

Milgram, S. (1963). Behavioural study of obedience. *Journal of Abnormal and Social Psychology,* 67:371–8.

Miller, G.A. (1956). The magical number seven, plus or minus two. *Psychological Review, 63,* 81–97.

Piaget, J. (1926). *The language and thought of the child.* London: Routledge & Kegan Paul.

Sheldon, W.H. (1954). *Atlas of men: A guide for somatotyping the adult male at all ages.* New York: Harper & Row.

Skinner, B.F. (1957). *Verbal behaviour of organisms.* New York: Appleton-Century-Crofts.

Sperry, R.W. (1964). The great cerebral commissure. *Scientific American,* 210: 42–52.

Thorndike, R.L. (1911). *Animal intelligence.* Macmillan.

Tinbergen, N. (1953). *Social Behaviour in Animals.* London: Methuen.

Toffler, A. (1970). *Future shock.* New York: Random House.

Tolman, E.C. (1932). *Purpose behaviour in animals and men.* New York: Appleton-Century-Crofts. (Reprint ed., 1967. New York: Irvington).

Torrance, E.P. (1972). Characteristics of creatively gifted children and youth. In Philip Trapp, E. and Himelstein, P. (eds), *The exceptional child.* New York: Appleton-Century-Crofts.

Treisman, A.M. (1964). Selective attention in man. *British Medical Bulletin,* 20: 12–16.

Vernon, M.D. (1960). *Intelligence and attainment tests.* London: University of London Press.